GLAUCUS

Copyright © 2006 BiblioBazaar
All rights reserved
ISBN: 1-4264-0318-6

CHARLES KINGSLEY

GLAUCUS

or
The Wonders of the Shore

Glaucus

DEDICATION.

MY DEAR MISS GRENFELL,

I CANNOT forego the pleasure of dedicating this little book to you; excepting of course the opening exhortation (needless enough in your case) to those who have not yet discovered the value of Natural History. Accept it as a memorial of pleasant hours spent by us already, and as an earnest, I trust, of pleasant hours to be spent hereafter (perhaps, too, beyond this life in the nobler world to come), in examining together the works of our Father in heaven.

Your grateful and faithful brother-in-law,
C. KINGSLEY.
BIDEFORD,
APRIL 24. 1855.

GLAUCUS;
OR, THE WONDERS OF THE SHORE.

You are going down, perhaps, by railway, to pass your usual six weeks at some watering-place along the coast, and as you roll along think more than once, and that not over-cheerfully, of what you shall do when you get there. You are half-tired, half-ashamed, of making one more in the ignoble army of idlers, who saunter about the cliffs, and sands, and quays; to whom every wharf is but a "wharf of Lethe," by which they rot "dull as the oozy weed." You foreknow your doom by sad experience. A great deal of dressing, a lounge in the club-room, a stare out of the window with the telescope, an attempt to take a bad sketch, a walk up one parade and down another, interminable reading of the silliest of novels, over which you fall asleep on a bench in the sun, and probably have your umbrella stolen; a purposeless fine-weather sail in a yacht, accompanied by many ineffectual attempts to catch a mackerel, and the consumption of many cigars; while your boys deafen your ears, and endanger your personal safety, by blazing away at innocent gulls and willocks, who go off to die slowly; a sport which you feel to be wanton, and cowardly, and cruel, and yet cannot find in your heart to stop, because "the lads have nothing else to do, and at all events it keeps them out of the billiard-room;" and after all, and worst of all, at night a soulless RECHAUFFE of third-rate London frivolity: this is the life-in-death in which thousands spend the golden weeks of summer, and in which you confess with a sigh that you are going to spend them.

Now I will not be so rude as to apply to you the old hymn-distich about one who

>" - finds some mischief still
> For idle hands to do:"

but does it not seem to you, that there must surely be many a thing worth looking at earnestly, and thinking over earnestly, in a world like this, about the making of the least part whereof God has employed ages and ages, further back than wisdom can guess or imagination picture, and upholds that least part every moment by laws and forces so complex and so wonderful, that science, when it tries to fathom them, can only learn how little it can learn? And does it not seem to you that six weeks' rest, free from the cares of town business and the whirlwind of town pleasure, could not be better spent than in examining those wonders a little, instead of wandering up and down like the many, still wrapt up each in his little world of vanity and self-interest, unconscious of what and where they really are, as they gaze lazily around at earth and sea and sky, and have

> "No speculation in those eyes
> Which they do glare withal"?

Why not, then, try to discover a few of the Wonders of the Shore? For wonders there are there around you at every step, stranger than ever opium-eater dreamed, and yet to be seen at no greater expense than a very little time and trouble.

Perhaps you smile, in answer, at the notion of becoming a "Naturalist:" and yet you cannot deny that there must be a fascination in the study of Natural History, though what it is is as yet unknown to you. Your daughters, perhaps, have been seized with the prevailing "Pteridomania," and are collecting and buying ferns, with Ward's cases wherein to keep them (for which you have to pay), and wrangling over unpronounceable names of species (which seem to he different in each new Fern-book that they buy), till the Pteridomania seems to you somewhat of a bore: and yet you cannot deny that they find an enjoyment in it, and are more active, more cheerful, more self-forgetful over it, than they would have been over novels and gossip, crochet and Berlin-wool. At least you will confess that the abomination of "Fancy-work" - that standing cloak for dreamy idleness (not to mention the injury which it does to poor starving needlewomen) - has all but vanished from your drawing-room since the "Lady-ferns" and "Venus's hair" appeared; and that you could not help yourself looking now and then at the said "Venus's hair," and agreeing that Nature's real beauties were somewhat superior to the ghastly woollen caricatures which they had superseded.

You cannot deny, I say, that there is a fascination in this same Natural History. For do not you, the London merchant, recollect how but last summer your douce and portly head-clerk was seized by two keepers in the act of wandering in Epping Forest at dead of night, with a dark lantern, a jar of strange sweet compound, and innumerable pocketfuls of pill-boxes; and found it very difficult to make either his captors or you believe that he was neither going to burn wheat-ricks, nor poison pheasants, but was simply "sugaring the trees for moths," as a blameless entomologist? And when, in self-justification, he took you to his house in Islington, and showed you the glazed and corked drawers full of delicate insects, which had evidently cost him in the collecting the spare hours of many busy years, and many a pound, too, out of his small salary, were you not a little puzzled to make out what spell there could be in those "useless" moths, to draw out of his warm bed, twenty miles down the Eastern Counties Railway, and into the damp forest like a deer-stealer, a sober white-headed Tim Linkinwater like him, your very best man of business, given to the reading of Scotch political economy, and gifted with peculiarly clear notions on the currency question?

It is puzzling, truly. I shall be very glad if these pages help you somewhat toward solving the puzzle.

We shall agree at least that the study of Natural History has become now-a-days an honourable one. A Cromarty stonemason was till lately - God rest his noble soul! - the most important man in the City of Edinburgh, by dint of a work on fossil fishes; and the successful investigator of the minutest animals takes place unquestioned among men of genius, and, like the philosopher of old Greece, is considered, by virtue of his science, fit company for dukes and princes. Nay, the study is now more than honourable; it is (what to many readers will be a far higher recommendation) even fashionable. Every well-educated person is eager to know something at least of the wonderful organic forms which surround him in every sunbeam and every pebble; and books of Natural History are finding their way more and more into drawing-rooms and school-rooms, and exciting greater thirst for a knowledge which, even twenty years ago, was considered superfluous for all but the professional student.

What a change from the temper of two generations since, when the naturalist was looked on as a harmless enthusiast, who went "bug- hunting," simply because he had not spirit to follow a fox! There are those alive who can recollect an amiable man being literally bullied out of the New Forest, because he dared to make a collection (at this moment, we believe, in some unknown abyss of that great Avernus, the British Museum) of fossil shells from those very Hordwell Cliffs, for exploring which there is now established a society of subscribers and correspondents. They can remember, too, when, on the first appearance of Bewick's "British Birds," the excellent sportsman who brought it down to the Forest was asked, Why on earth he had bought a book about "cock sparrows"? and had to justify himself again and again, simply by lending the book to his brother sportsmen, to convince them that there were rather more than a dozen sorts of birds (as they then held) indigenous to Hampshire. But the book, perhaps, which turned the tide in favour of Natural History, among the higher classes at least, in the south of England, was White's "History of Selborne." A Hampshire gentleman and sportsman, whom everybody knew, had taken the trouble to write a book about the birds and the weeds in his own parish, and the every-day things which went on under his eyes, and everyone else's. And all gentlemen, from the Weald of Kent to the Vale of Blackmore, shrugged their shoulders mysteriously, and said, "Poor fellow!" till they opened the book itself, and discovered to their surprise that it read like any novel. And then came a burst of confused, but honest admiration; from the young squire's "Bless me! who would have thought that there were so many wonderful things to be seen in one's own park!" to the old squire's more morally valuable "Bless me! why, I have seen that and that a hundred times, and never thought till now how wonderful they were!"

There were great excuses, though, of old, for the contempt in which the naturalist was held; great excuses for the pitying tone of banter with which the Spectator talks of "the ingenious" Don Saltero (as no doubt the Neapolitan gentleman talked of Ferrante

Imperato the apothecary, and his museum); great excuses for Voltaire, when he classes the collection of butterflies among the other "bizarreries de l'esprit humain." For, in the last generation, the needs of the world were different. It had no time for butterflies and fossils. While Buonaparte was hovering on the Boulogne coast, the pursuits and the education which were needed were such as would raise up men to fight him; so the coarse, fierce, hard-handed training of our grandfathers came when it was wanted, and did the work which was required of it, else we had not been here now. Let us be thankful that we have had leisure for science; and show now in war that our science has at least not unmanned us.

Moreover, Natural History, if not fifty years ago, certainly a hundred years ago, was hardly worthy of men of practical common sense. After, indeed, Linne, by his invention of generic and specific names, had made classification possible, and by his own enormous labours had shown how much could be done when once a method was established, the science has grown rapidly enough. But before him little or nothing had been put into form definite enough to allure those who (as the many always will) prefer to profit by others' discoveries, than to discover for themselves; and Natural History was attractive only to a few earnest seekers, who found too much trouble in disencumbering their own minds of the dreams of bygone generations (whether facts, like cockatrices, basilisks, and krakens, the breeding of bees out of a dead ox, and of geese from barnacles; or theories, like those of elements, the VIS PLASTRIX in Nature, animal spirits, and the other musty heirlooms of Aristotleism and Neo-platonism), to try to make a science popular, which as yet was not even a science at all. Honour to them, nevertheless. Honour to Ray and his illustrious contemporaries in Holland and France. Honour to Seba and Aldrovandus; to Pomet, with his "Historie of Drugges;" even to the ingenious Don Saltero, and his tavern-museum in Cheyne Walk. Where all was chaos, every man was useful who could contribute a single spot of organized standing ground in the shape of a fact or a specimen. But it is a question whether Natural History would have ever attained its present honours, had not Geology arisen, to connect every other branch of Natural History with problems as vast and awful as they are captivating to the imagination. Nay, the very opposition with which Geology met was of as great benefit to the sister sciences as to itself. For, when questions belonging to the most sacred hereditary beliefs of Christendom were supposed to be affected by the verification of a fossil shell, or the proving that the Maestricht "homo diluvii testis" was, after all, a monstrous eft, it became necessary to work upon Conchology, Botany, and Comparative Anatomy, with a care and a reverence, a caution and a severe induction, which had been never before applied to them; and thus gradually, in the last half-century, the whole choir of cosmical sciences have acquired a soundness, severity, and fulness, which render them, as mere intellectual exercises, as valuable to a manly mind as Mathematics and Metaphysics.

But how very lately have they attained that firm and honourable standing ground! It is a question whether, even twenty years ago, Geology, as it then stood, was worth troubling one's head about, so little had been really proved. And heavy and uphill was the work, even within the last fifteen years, of those who stedfastly set themselves to the task of proving and of asserting at all risks, that the Maker of the coal seam and the diluvial cave could not be a "Deus quidam deceptor," and that the facts which the rock and the silt revealed were sacred, not to be warped or trifled with for the sake of any cowardly and hasty notion that they contradicted His other messages. When a few more years are past, Buckland and Sedgwick, Murchison and Lyell, Delabche and Phillips, Forbes and Jamieson, and the group of brave men who accompanied and followed them, will be looked back to as moral benefactors of their race; and almost as martyrs, also, when it is remembered how much misunderstanding, obloquy, and plausible folly they had to endure from well-meaning fanatics like Fairholme or Granville Penn, and the respectable mob at their heels who tried (as is the fashion in such cases) to make a hollow compromise between fact and the Bible, by twisting facts just enough to make them fit the fancied meaning of the Bible, and the Bible just enough to make it fit the fancied meaning of the facts. But there were a few who would have no compromise; who laboured on with a noble recklessness, determined to speak the thing which they had seen, and neither more nor less, sure that God could take better care than they of His own everlasting truth. And now they have conquered: the facts which were twenty years ago denounced as contrary to Revelation, are at last accepted not merely as consonant with, but as corroborative thereof; and sound practical geologists - like Hugh Miller, in his "Footprints of the Creator," and Professor Sedgwick, in the invaluable notes to his "Discourse on the Studies of Cambridge" - have wielded in defence of Christianity the very science which was faithlessly and cowardly expected to subvert it.

But if you seek, reader, rather for pleasure than for wisdom, you can find it in such studies, pure and undefiled.

Happy, truly, is the naturalist. He has no time for melancholy dreams. The earth becomes to him transparent; everywhere he sees significancies, harmonies, laws, chains of cause and effect endlessly interlinked, which draw him out of the narrow sphere of self-interest and self-pleasing, into a pure and wholesome region of solemn joy and wonder. He goes up some Snowdon valley; to him it is a solemn spot (though unnoticed by his companions), where the stag's-horn clubmoss ceases to straggle across the turf, and the tufted alpine clubmoss takes its place: for he is now in a new world; a region whose climate is eternally influenced by some fresh law (after which he vainly guesses with a sigh at his own ignorance), which renders life impossible to one species, possible to another. And it is a still more solemn thought to him, that it was not always so; that aeons and ages back, that rock which he passed a thousand feet below was fringed, not as now with fern and blue bugle, and white bramble-flowers, but perhaps with the alp- rose and the "gemsen-kraut" of Mont Blanc, at least with Alpine Saxifrages which have now retreated a thousand feet up the mountain side, and with the blue Snow-Gentian, and the Canadian Sedum, which have all but vanished out of the British Isles. And what is it which tells him that strange story? Yon smooth and rounded surface of rock, polished, remark, across the strata and against the grain; and furrowed here and there, as if by iron talons, with long parallel scratches. It was the crawling of a glacier which polished that rock-face; the stones fallen from Snowdon peak into the half-liquid lake of ice above, which ploughed those furrows. Æons and aeons ago, before the time when Adam first

> "Embraced his Eve in happy hour,
> And every bird in Eden burst
> In carol, every bud in flower,"

those marks were there; the records of the "Age of ice;" slight, truly; to be effaced by the next farmer who needs to build a wall; but unmistakeable, boundless in significance, like Crusoe's one savage footprint on the sea-shore; and the naturalist acknowledges the finger-mark of God, and wonders, and worships.

Happy, especially, is the sportsman who is also a naturalist: for as he roves in pursuit of his game, over hills or up the beds of streams where no one but a sportsman ever thinks of going, he will be certain to see things noteworthy, which the mere naturalist would never find, simply because he could never guess that they were there to be found. I do not speak merely of the rare birds which may be shot, the curious facts as to the habits of fish which may be observed, great as these pleasures are. I speak of the scenery, the weather, the geological formation of the country, its vegetation, and the living habits of its denizens. A sportsman, out in all weathers, and often dependent for success on his knowledge of "what the sky is going to do," has opportunities for becoming a meteorologist which no one beside but a sailor possesses; and one has often longed for a scientific gamekeeper or huntsman, who, by discovering a law for the mysterious and seemingly capricious phenomena of "scent," might perhaps throw light on a hundred dark passages of hygrometry. The fisherman, too, - what an inexhaustible treasury of wonder lies at his feet, in the subaqueous world of the commonest mountain burn! All the laws which mould a world are there busy, if he but knew it, fattening his trout for him, and making them rise to the fly, by strange electric influences, at one hour rather than at another.

Many a good geognostic lesson, too, both as to the nature of a country's rocks, and as to the laws by which strata are deposited, may an observing man learn as he wades up the bed of a trout-stream; not to mention the strange forms and habits of the tribes of water-insects. Moreover, no good fisherman but knows, to his sorrow, that there are plenty of minutes, ay, hours, in each day's fishing in which he would be right glad of any employment better than trying to

"Call spirits from the vasty deep,"

who will not

"Come when you do call for them."

What to do, then? You are sitting, perhaps, in your coracle, upon some mountain tarn, waiting for a wind, and waiting in vain.

"Keine luft an keine seite,
Todes-stille forchterlich;"

as Gthe has it -

"Und der schiffer sieht bekommert
Glatte flche rings umher."

You paddle to the shore on the side whence the wind ought to come, if it had any spirit in it; tie the coracle to a stone, light your cigar, lie down on your back upon the grass, grumble, and finally fall asleep. In the meanwhile, probably, the breeze has come on, and there has been half-an-hour's lively fishing curl; and you wake just in time to see the last ripple of it sneaking off at the other side of the lake, leaving all as dead-calm as before.

Now how much better, instead of falling asleep, to have walked quietly round the lake side, and asked of your own brains and of Nature the question, "How did this lake come here? What does it mean?"

It is a hole in the earth. True, but how was the hole made? There must have been huge forces at work to form such a chasm. Probably the mountain was actually opened from within by an earthquake; and when the strata fell together again, the portion at either end of the chasm, being perhaps crushed together with greater force, remained higher than the centre, and so the water lodged between them. Perhaps it was formed thus. You will at least agree that its formation must have been a grand sight enough, and one during which a spectator would have had some difficulty in keeping his footing.

And when you learn that this convulsion probably took plus at the bottom of an ocean hundreds of thousands of years ago, you have at least a few thoughts over which to ruminate, which will make you at once too busy to grumble, and ashamed to grumble.

17

Yet, after all, I hardly think the lake was formed in this way, and suspect that it may have been dry for ages after it emerged from the primeval waves, and Snowdonia was a palm-fringed island in a tropic sea. Let us look the place over more fully.

You see the lake is nearly circular; on the side where we stand the pebbly beach is not six feet above the water, and slopes away steeply into the valley behind us, while before us it shelves gradually into the lake; forty yards out, as you know, there is not ten feet water; and then a steep bank, the edge whereof we and the big trout know well, sinks suddenly to unknown depths. On the opposite side, that flat-topped wall of rock towers up shoreless into the sky, seven hundred feet perpendicular; the deepest water of all we know is at its very foot. Right and left, two shoulders of down slope into the lake. Now turn round and look down the gorge. Remark that this pebble bank on which we stand reaches some fifty yards downward: you see the loose stones peeping out everywhere. We may fairly suppose that we stand on a dam of loose stones, a hundred feet deep.

But why loose stones? - and if so, what matter? and what wonder? There are rocks cropping out everywhere down the hillside.

Because if you will take up one of these stones and crack it across, you will see that it is not of the same stuff as those said rocks. Step into the next field and see. That rock is the common Snowdon slate, which we see everywhere. The two shoulders of down, right and left, are slate, too; you can see that at a glance. But the stones of the pebble bank are a close-grained, yellow-spotted rock. They are Syenite; and (you may believe me or not, as you will) they were once upon a time in the condition of a hasty pudding heated to some 800 degrees of Fahrenheit, and in that condition shoved their way up somewhere or other through these slates. But where? whence on earth did these Syenite pebbles come?

Let us walk round to the cliff on the opposite side and see. It is worth while; for even if my guess be wrong, there is good spinning with a brass minnow round the angles of the rocks.

Now see. Between the cliff-foot and the sloping down is a crack, ending in a gully; the nearer side is of slate, and the further side, the cliff itself, is - why, the whole cliff is composed of the very same stone as the pebble ridge.

Now, my good friend, how did these pebbles get three hundred yards across the lake? Hundreds of tons, some of them three feet long: who carried them across? The old Cymry were not likely to amuse themselves by making such a breakwater up here in No-man's-land, two thousand feet above the sea: but somebody or something must have carried them; for stones do not fly, nor swim either.

Shot out of a volcano? As you seem determined to have a prodigy, it may as well be a sufficiently huge one.

Well - these stones lie altogether; and a volcano would have hardly made so compact a shot, not being in the habit of using Eley's wire cartridges. Our next hope of a solution lies in John Jones, who carried up the coracle. Hail him, and ask him what is on the top of that cliff . . . So, "Plainshe and pogshe, and another Llyn."

Very good. Now, does it not strike you that this whole cliff has a remarkably smooth and plastered look, like a hare's run up an earthbank? And do you not see that it is polished thus only over the lake? that as soon as the cliff abuts on the downs right and left, it forms pinnacles, caves, broken angular boulders? Syenite usually does so in our damp climate, from the "weathering" effect of frost and rain: why has it not done so over the lake? On that part something (giants perhaps) has been scrambling up or down on a very large scale, and so rubbed off every corner which was inclined to come away, till the solid core of the rock was bared. And may not those mysterious giants have had a hand in carrying the stones across the lake? . . . Really, I am not altogether jesting. Think a while what agent could possibly have produced either one or both of these effects?

There is but one; and that, if you have been an Alpine traveller - much more if you have been a Chamois hunter - you have seen many a time (whether you knew it or not) at the very same work.

Ice? Yes; ice; Hrymir the frost-giant, and no one else. And if you will look at the facts, you will see how ice may have done it.

Our friend John Jones's report of plains and bogs and a lake above makes it quite possible that in the "Ice age" (Glacial Epoch, as the big-word-mongers call it) there was above that cliff a great neve, or snowfield, such as you have seen often in the Alps at the head of each glacier. Over the face of this cliff a glacier has crawled down from that neve, polishing the face of the rock in its descent: but the snow, having no large and deep outlet, has not slid down in a sufficient stream to reach the vale below, and form a glacier of the first order; and has therefore stopped short on the other side of the lake, as a glacier of the second order, which ends in an ice-cliff hanging high up on the mountain side, and kept from further progress by daily melting. If you have ever gone up the Mer de Glace to the Tacul, you saw a magnificent specimen of this sort on your right hand, just opposite the Tacul, in the Glacier de Trelaporte, which comes down from the Aiguille de Charmoz.

This explains our pebble-ridge. The stones which the glacier rubbed off the cliff beneath it it carried forward, slowly but surely, till they saw the light again in the face of the ice-cliff, and dropped out of it under the melting of the summer sun, to form a huge dam across the ravine; till, the "Ice age" past, a more genial climate succeeded, and neve and glacier melted away: but the "moraine" of stones did not, and remains to this day, as the dam which keeps up the waters of the lake.

There is my explanation. If you can find a better, do: but remember always that it must include an answer to - "How did the stones get across the lake?"

Now, reader, we have had no abstruse science here, no long words, not even a microscope or a book: and yet we, as two plain sportsmen, have gone back, or been led back by fact and common sense, into the most awful and sublime depths, into an epos of the destruction and re-creation of a former world.

This is but a single instance; I might give hundreds. This one, nevertheless, may have some effect in awakening you to the boundless world of wonders which is all around you, and make you ask yourself seriously, "What branch of Natural History shall I begin to investigate, if it be but for a few weeks, this summer?"

To which I answer, Try "the Wonders of the Shore." There are along every sea-beach more strange things to be seen, and those to be seen easily, than in any other field of observation which you will find in these islands. And on the shore only will you have the enjoyment of finding new species, of adding your mite to the treasures of science.

For not only the English ferns, but the natural history of all our land species, are now well-nigh exhausted. Our home botanists and ornithologists are spending their time now, perforce, in verifying a few obscure species, and bemoaning themselves, like Alexander, that there are no more worlds left to conquer. For the geologist, indeed, and the entomologist, especially in the remoter districts, much remains to be done, but only at a heavy outlay of time, labour, and study; and the dilettante (and it is for dilettanti, like myself, that I principally write) must be content to tread in the tracks of greater men who have preceded him, and accept at second or third hand their foregone conclusions.

But this is most unsatisfactory; for in giving up discovery, one gives up one of the highest enjoyments of Natural History. There is a mysterious delight in the discovery of a new species, akin to that of seeing for the first time, in their native haunts, plants or animals of which one has till then only read. Some, surely, who read these pages have experienced that latter delight; and, though they might find it hard to define whence the pleasure arose, know well that it was a solid pleasure, the memory of which they would not give up for hard cash. Some, surely, can recollect, at their first sight of the Alpine Soldanella, the Rhododendron, or the black Orchis, growing upon the edge of the eternal snow, a thrill of emotion not unmixed with awe; a sense that they were, as it were, brought face to face with the creatures of another world; that Nature was independent of them, not merely they of her; that trees were not merely made to build their houses, or herbs to feed their cattle, as they looked on those wild gardens amid the wreaths of the untrodden snow, which had lifted their gay flowers to the sun year after year since the foundation of the world, taking no heed of man, and all the coil which he keeps in the valleys far below.

And even, to take a simpler instance, there are those who will excuse, or even approve of, a writer for saying that, among the memories of a month's eventful tour, those which stand out as beacon-points, those round which all the others group themselves, are the first wolf-track by the road-side in the Kyllwald; the first sight of the blue and green Roller-birds, walking behind the plough like rooks in the tobacco-fields of Wittlich; the first ball of Olivine scraped out of the volcanic slag-heaps of the Dreisser- Weiher; the first pair of the Lesser Bustard flushed upon the downs of the Mosel-kopf; the first sight of the cloud of white Ephemerae, fluttering in the dusk like a summer snowstorm between us and the black cliffs of the Rheinstein, while the broad Rhine beneath flashed blood-red in the blaze of the lightning and the fires of the Mausenthurm - a lurid Acheron above which seemed to hover ten thousand unburied ghosts; and last, but not least, on the lip of the vast Mosel-kopf crater - just above the point where the weight of the fiery lake has burst the side of the great slag-cup, and rushed forth between two cliffs of clink-stone across the downs, in a clanging stream of fire, damming up rivulets, and blasting its path through forests, far away toward the valley of the Moselle - the sight of an object for which was forgotten for the moment that battle-field of the Titans at our feet, and the glorious panorama, Hundsruck and Taunus, Siebengebirge and Ardennes, and all the crater peaks around; and which was - smile not, reader - our first yellow foxglove.

But what is even this to the delight of finding a new species? - of rescuing (as it seems to you) one more thought of the Divine mind from Hela, and the realms of the unknown, unclassified, uncomprehended? As it seems to you: though in reality it only seems so, in a world wherein not a sparrow falls to the ground unnoticed by our Father who is in heaven.

The truth is, the pleasure of finding new species is too great; it is morally dangerous; for it brings with it the temptation to look on the thing found as your own possession, all but your own creation; to pride yourself on it, as if God had not known it for ages since; even to squabble jealously for the right of having it named after you, and of being recorded in the Transactions of I- know-not-what Society as its first discoverer:- as if all the angels in heaven had not been admiring it, long before you were born or thought of.

But to be forewarned is to be forearmed; and I seriously counsel you to try if you cannot find something new this summer along the coast to which you are going. There is no reason why you should not be so successful as a friend of mine who, with a very slight smattering of science, and very desultory research, obtained in one winter from the Torbay shores three entirely new species, beside several rare animals which had escaped all naturalists since the lynx-eye of Colonel Montagu discerned them forty years ago.

And do not despise the creatures because they are minute. No doubt we should most of us prefer discovering monstrous apes in the tropical forests of Borneo, or stumbling upon herds of gigantic Ammon sheep amid the rhododendron thickets of the Himalaya: but it cannot be; and "he is a fool," says old Hesiod, "who knows not how much better half is than the whole." Let us be content with what is within our reach. And doubt not that in these tiny creatures are mysteries more than we shall ever fathom.

The zoophytes and microscopic animalcules which people every shore and every drop of water, have been now raised to a rank in the human mind more important, perhaps, than even those gigantic monsters whose models fill the lake at the Crystal Palace. The research which has been bestowed, for the last century, upon these once unnoticed atomies has well repaid itself; for from no branch of physical science has more been learnt of the SCIENTIA SCIENTIARUM, the priceless art of learning; no branch of science has more utterly confounded a wisdom of the wise, shattered to pieces systems and theories, and the idolatry of arbitrary names, and taught man to be silent while his Maker speaks, than this apparent pedantry of zoophytology, in which our old distinctions of "animal," "vegetable," and "mineral" are trembling in the balance, seemingly ready to vanish like their fellows - "the four elements" of fire, earth, air, and water. No branch of science has helped so much to sweep away that sensuous idolatry of mere size, which tempts man to admire and respect objects in proportion to the number of feet or inches which they occupy in space. No branch of science, moreover, has been more humbling to the boasted rapidity and omnipotence of the human reason, or has more taught those who have eyes to see, and hearts to understand, how weak and wayward, staggering and slow, are the steps of our fallen race (rapid and triumphant enough in that broad road of theories which leads to intellectual destruction) whensoever they tread the narrow path of true science, which leads (if I may be allowed to transfer our Lord's great parable from moral to intellectual matters) to Life; to the living and permanent knowledge of living things and of the laws of their existence. Humbling, truly, to one who looks back to the summer of 1754, when good Mr. Ellis, the wise and benevolent West Indian merchant, read before the Royal Society his paper proving the animal nature of corals, and followed it up the year after by that "Essay toward a Natural History of the Corallines, and other like Marine Productions of the British Coasts," which forms the groundwork of all our knowledge on the subject to this day. The chapter in Dr. G. Johnston's "British Zoophytes," p. 407, or the excellent little RESUME thereof in Dr. Landsborough's book on the same subject, is really a saddening one, as one sees how loth were, not merely dreamers like, Marsigli or Bonnet, but sound-headed men like Pallas and Linne, to give up the old sense-bound fancy, that these corals were vegetables, and their polypes some sort of living flowers. Yet, after all, there are excuses for them. Without our improved microscopes, and while the sciences of comparative anatomy and chemistry were yet infantile, it was difficult to believe what was the truth; and for this simple reason: that, as usual, the truth, when discovered, turned out far more startling and prodigious than the dreams which men had hastily substituted for it; more strange than Ovid's old story that the coral was soft under the sea, and hardened by exposure to air; than Marsigli's notion, that the coral-polypes were its flowers; than Dr. Parsons' contemptuous denial, that these complicated forms could be "the operations of little, poor, helpless, jelly-like animals, and not the work of more sure vegetation;" than Baker the microscopist's detailed theory of their being produced by the crystallization of the mineral salts in the sea-water, just as he had seen "the particles of mercury and copper in aquafortis assume tree-like forms, or curious delineations of mosses and minute shrubs on slates and stones, owing to the shooting of salts intermixed with mineral particles:" - one smiles at it now: yet these men were no less sensible than we; and if we know better, it is only because other men, and those few and far between, have laboured amid disbelief, ridicule, and error; needing again and again to retrace their steps, and to unlearn more than they learnt, seeming to go backwards when they were really progressing most:

and now we have entered into their labours, and find them, as I have just said, more wondrous than all the poetic dreams of a Bonnet or a Darwin. For who, after all, to take a few broad instances (not to enlarge on the great root-wonder of a number of distinct individuals connected by a common life, and forming a seeming plant invariable in each species), would have dreamed of the "bizarreries" which these very zoophytes present in their classification?

You go down to any shore after a gale of wind, and pick up a few delicate little sea-ferns. You have two in your hand, which probably look to you, even under a good pocket magnifier, identical or nearly so. [1] But you are told to your surprise, that however like the dead horny polypidoms which you hold may be, the two species of animal which have formed them are at least as far apart in the scale of creation as a quadruped is from a fish. You see in some Musselburgh dredger's boat the phosphorescent sea-pen (unknown in England), a living feather, of the look and consistency of a cock's comb; or the still stranger sea-rush (VIRGULARIA MIRABILIS), a spine a foot long, with hundreds of rosy flowerets arranged in half-rings round it from end to end; and you are told that these are the congeners of the great stony Venus's fan which hangs in seamen's cottages, brought home from the West Indies. And ere you have done wondering, you hear that all three are congeners of the ugly, shapeless, white "dead man's hand," which you may pick up after a storm on any shore. You have a beautiful madrepore or brain-stone on your mantel-piece, brought home from some Pacific coral-reef. You are to believe that its first cousins are the soft, slimy sea-anemones which you see expanding their living flowers in every rock-pool - bags of sea-water, without a trace of bone or stone. You must believe it; for in science, as in higher matters, he who will walk surely, must "walk by faith and not by sight."

These are but a few of the wonders which the classification of marine animals affords; and only drawn from one class of them, though almost as common among every other family of that submarine world whereof Spenser sang -

"Oh, what an endless work have I in hand,
To count the sea's abundant progeny!
Whose fruitful seed far passeth those in land,
And also those which won in th' azure sky,
For much more earth to tell the stars on high,
Albe they endless seem in estimation,
Than to recount the sea's posterity;
So fertile be the flouds in generation,
So huge their numbers, and so numberless their nation."

But these few examples will be sufficient to account both for the slow pace at which the knowledge of sea-animals has progressed, and for the allurement which men of the highest attainments have found, and still find, in it. And when to this we add the marvels which meet us at every step in the anatomy and the reproduction of these creatures, and in the chemical and mechanical functions which they fulfil in the great economy of our planet, we cannot wonder at finding that books which treat of them carry with them a certain charm of romance, and feed the play of fancy, and that love of the marvellous which is inherent in man, at the same time that they lead the reader to more solemn and lofty trains of thought, which can find their full satisfaction only in self-forgetful worship, and that hymn of praise which goes up ever from land and sea, as well as from saints and martyrs and the heavenly host, "O all ye works of the Lord, and ye, too, spirits and souls of the righteous, praise Him, and magnify Him for ever!"

I have said, that there were excuses for the old contempt of the study of Natural History. I have said, too, it may be hoped, enough to show that contempt to be now ill-founded. But still, there are those who regard it as a mere amusement, and that as a somewhat effeminate one; and think that it can at best help to while away a leisure hour harmlessly, and perhaps usefully, as a substitute for coarser sports, or for the reading of novels.

Those, however, who have followed it out, especially on the sea-shore, know better. They can tell from experience, that over and above its accessory charms of pure sea-breezes, and wild rambles by cliff and loch, the study itself has had a weighty moral effect upon their hearts and spirits. There are those who can well understand how the good and wise John Ellis, amid all his philanthropic labours for the good of the West Indies, while he was spending his intellect and fortune in introducing into our tropic settlements the bread-fruit, the mangosteen, and every plant and seed which he hoped might be useful for medicine, agriculture, and commerce, could yet feel himself justified in devoting large portions of his ever well-spent time to the fighting the battle of the corallines against Parsons and the rest, and even in measuring pens with Linne, the prince of naturalists.

There are those who can sympathise with the gallant old Scotch officer mentioned by some writer on sea-weeds, who, desperately wounded in the breach at Badajos, and a sharer in all the toils and triumphs of the Peninsular war, could in his old age show a rare sea-weed with as much triumph as his well-earned medals, and talk over a tiny spore-capsule with as much zest as the records of sieges and battles. Why not? That temper which made him a good soldier may very well have made him a good naturalist also. The late illustrious geologist, Sir Roderick Murchison, was also an old Peninsular officer. I doubt not that with him, too, the experiences of war may have helped to fit him for the studies of peace. Certainly, the best naturalist, as far as logical acumen, as well as earnest research, is concerned, whom England has ever seen, was the Devonshire squire, Colonel George Montagu, of whom the late E. Forbes well says, that "had he been educated a physiologist" (and not, as he was, a soldier and a sportsman), "and made the study of Nature his aim and not his amusement, his would have been one of the greatest names in the whole range of British science." I question, nevertheless, whether he would not have lost more than he would have gained by a different training. It might have made him a more learned systematizer; but would it have quickened in him that "seeing" eye of the true soldier and sportsman, which makes Montagu's descriptions indelible word- pictures, instinct with life and truth? "There is no question," says E. Forbes, after bewailing the vagueness of most naturalists, "about the identity of any animal Montagu described ... He was a forward-looking philosopher; he spoke of every creature as if one exceeding like it, yet different from it, would be washed up by the waves next tide. Consequently his descriptions are permanent."

Scientific men will recognize in this the highest praise which can be bestowed, because it attributes to him the highest faculty - The Art of Seeing; but the study and the book would not have given that. It is God's gift wheresoever educated: but its true school- room is the camp and the ocean, the prairie and the forest; active, self-helping life, which can grapple with Nature herself: not merely with printed-books about her. Let no one think that this same Natural History is a pursuit fitted only for effeminate or pedantic men. I should say, rather, that the qualifications required for a perfect naturalist are as many and as lofty as were required, by old chivalrous writers, for the perfect knight-errant of the Middle Ages: for (to sketch an ideal, of which I am happy to say our race now affords many a fair realization) our perfect naturalist should be strong in body; able to haul a dredge, climb a rock, turn a boulder, walk all day, uncertain where he shall eat or rest; ready to face sun and rain, wind and frost, and to eat or drink thankfully anything, however coarse or meagre; he should know how to swim for his life, to pull an oar, sail a boat, and ride the first horse which comes to hand; and, finally, he should be a thoroughly good shot, and a skilful fisherman; and, if he go far abroad, be able on occasion to fight for his life.

For his moral character, he must, like a knight of old, be first of all gentle and courteous, ready and able to ingratiate himself with the poor, the ignorant, and the savage; not only because foreign travel will be often otherwise impossible, but because he knows how much invaluable local information can be only obtained from fishermen, miners, hunters, and tillers of the soil. Next, he should be brave and enterprising, and withal patient and undaunted; not merely in travel, but in investigation; knowing (as Lord Bacon might have put it) that the kingdom of Nature, like the kingdom of heaven, must be taken by violence, and that only to those who knock long and earnestly does the great mother open the doors of her sanctuary. He must be of a reverent turn of mind also; not rashly discrediting any reports, however vague and fragmentary; giving man credit always for some germ of truth, and giving Nature credit for an inexhaustible fertility and variety, which will keep him his life long always reverent, yet never superstitious; wondering at the commonest, but not surprised by the most strange; free from the idols of size and sensuous loveliness; able to see grandeur in the minutest objects, beauty, in the most ungainly; estimating each thing not carnally, as the vulgar do, by its size or its pleasantness to the senses, but spiritually, by the amount of Divine thought revealed to Man therein; holding every phenomenon worth the noting down; believing that every pebble holds a treasure, every bud a revelation; making it a point of conscience to pass over nothing through laziness or hastiness, lest the vision once offered and despised should be withdrawn; and looking at every object as if he were never to behold it again.

Moreover, he must keep himself free from all those perturbations of mind which not only weaken energy, but darken and confuse the inductive faculty; from haste and laziness, from melancholy, testiness, pride, and all the passions which make men see only what they wish to see. Of solemn and scrupulous reverence for truth; of the habit of mind which regards each fact and discovery, not as our own possession, but as the possession of its Creator, independent of us, our tastes, our needs, or our vainglory, I hardly need to speak; for it is the very essence of a nature's faculty - the very tenure of his existence: and without truthfulness science would be as impossible now as chivalry would have been of old.

And last, but not least, the perfect naturalist should have in him the very essence of true chivalry, namely, self-devotion; the desire to advance, not himself and his own fame or wealth, but knowledge and mankind. He should have this great virtue; and in spite of many shortcomings (for what man is there who liveth and sinneth not?), naturalists as a class have it to a degree which makes them stand out most honourably in the midst of a self-seeking and mammonite generation, inclined to value everything by its money price, its private utility. The spirit which gives freely, because it knows that it has received freely; which communicates knowledge without hope of reward, without jealousy and rivalry, to fellow- students and to the world; which is content to delve and toil comparatively unknown, that from its obscure and seemingly worthless results others may derive pleasure, and even build up great fortunes, and change the very face of cities and lands, by the practical use of some stray talisman which the poor student has invented in his laboratory; - this is the spirit which is abroad among our scientific men, to a greater degree than it ever has been among any body of men for many a century past; and might well be copied by those who profess deeper purposes and a more exalted calling, than the discovery of a new zoophyte, or the classification of a moorland crag.

And it is these qualities, however imperfectly they may be realized in any individual instance, which make our scientific men, as a class, the wholesomest and pleasantest of companions abroad, and at home the most blameless, simple, and cheerful, in all domestic relations; men for the most part of manful heads, and yet of childlike hearts, who have turned to quiet study, in these late piping times of peace, an intellectual health and courage which might have made them, in more fierce and troublous times, capable of doing good service with very different instruments than the scalpel and the microscope.

I have been sketching an ideal: but one which I seriously recommend to the consideration of all parents; for, though it be impossible and absurd to wish that every young man should grow up a naturalist by profession, yet this age offers no more wholesome training, both moral and intellectual, than that which is given by instilling into the young an early taste for outdoor physical science. The education of our children is now more than ever a puzzling problem, if by education we mean the development of the whole humanity, not merely of some arbitrarily chosen part of it.

How to feed the imagination with wholesome food, and teach it to despise French novels, and that sugared slough of sentimental poetry, in comparison with which the old fairy-tales and ballads were manful and rational; how to counteract the tendency to shallowed and conceited sciolism, engendered by hearing popular lectures on all manner of subjects, which can only be really learnt by stern methodic study; how to give habits of enterprise, patience, accurate observation, which the counting-house or the library will never bestow; above all, how to develop the physical powers, without engendering brutality and coarseness - are questions becoming daily more and more puzzling, while they need daily more and more to be solved, in an age of enterprise, travel, and emigration, like the present. For the truth must be told, that the great majority of men who are now distinguished by commercial success, have had a training the directly opposite to that which they are giving to their sons. They are for the most part men who have migrated from the country to the town, and had in their youth all the advantages of a sturdy and manful hill-side or sea-side training; men whose bodies were developed, and their lungs fed on pure breezes, long before they brought to work in the city the bodily and mental strength which they had gained by loch and moor.

But it is not so with their sons. Their business habits are learnt in the counting-house; a good school, doubtless, as far as it goes: but one which will expand none but the lowest intellectual faculties; which will make them accurate accountants, shrewd computers and competitors, but never the originators of daring schemes, men able and willing to go forth to replenish the earth and subdue it. And in the hours of relaxation, how much of their time is thrown away, for want of anything better, on frivolity, not to say on secret profligacy, parents know too well; and often shut their eyes in very despair to evils which they know not how to cure. A frightful majority of our middle-class young men are growing up effeminate, empty of all knowledge but what tends directly to the making of a fortune; or rather, to speak correctly, to the keeping up the fortunes which their fathers have made for them; while of the minority, who are indeed thinkers and readers, how many women as well as men have we seen wearying their souls with study undirected, often misdirected; craving to learn, yet not knowing how or what to learn; cultivating, with unwholesome energy, the head at the expense of the body and the heart; catching up with the most capricious self-will one mania after another, and tossing it away again for some new phantom; gorging the memory with facts which no one has taught them to arrange, and the reason with problems which they have no method for solving; till they fret themselves in a chronic fever of the brain, which too often urge them on to plunge, as it were, to cool the inward fire, into the ever-restless seas of doubt or of superstition. It is a sad picture. There are many who may read these pages whose hearts will tell them that it is a true one. What is wanted in these cases is a methodic and scientific habit of mind; and a class of objects on which to exercise that habit, which will fever neither the speculative intellect nor the moral sense; and those physical science will give, as nothing else can give it.

Moreover, to revert to another point which we touched just now, man has a body as well as a mind; and with the vast majority there will be no MENS SANA unless there be a CORPUS SANUM for it to inhabit.

And what outdoor training to give our youths is, as we have already said, more than ever puzzling. This difficulty is felt, perhaps, less in Scotland than in England. The Scotch climate compels hardiness; the Scotch bodily strength makes it easy; and Scotland, with her mountain-tours in summer, and her frozen lochs in winter, her labyrinth of sea-shore, and, above all, that priceless boon which Providence has bestowed on her, in the contiguity of her great cities to the loveliest scenery, and the hills where every breeze is health, affords facilities for healthy physical life unknown to the Englishman, who has no Arthur's Seat towering above his London, no Western Islands sporting the ocean firths beside his Manchester. Field sports, with the invaluable training which they give, if not

> "The reason firm,"

yet still

> "The temperate will,
> Endurance, foresight, strength, and skill,"

have become impossible for the greater number: and athletic exercises are now, in England at least, becoming more and more artificialized and expensive; and are confined more and more - with the honourable exception of the football games in Battersea Park - to our Public Schools and the two elder Universities. All honour, meanwhile, to the Volunteer movement, and its moral as well as its physical effects. But it is only a comparatively few of the very sturdiest who are likely to become effective Volunteers, and so really gain the benefits of learning to be soldiers. And yet the young man who has had no substitute for such occupations will cut but a sorry figure in Australia, Canada, or India; and if he stays at home, will spend many a pound in doctors' bills, which could have been better employed elsewhere. "Taking a walk" - as one would take a pill or a draught - seems likely soon to become the only form of outdoor existence possible for too many inhabitants of the British Isles. But a walk without an object, unless in the most lovely and novel of scenery, is a poor exercise; and as a recreation, utterly nil. I never knew two young lads go out for a "constitutional," who did not, if they were commonplace youths, gossip the whole way about things better left unspoken; or, if they were clever ones, fall on arguing and brainsbeating on politics or metaphysics from the moment they left the door, and return with their wits even more heated and tired than they were when they set out. I cannot help fancying that Milton made a mistake in a certain celebrated passage; and that it was not "sitting on a hill apart," but tramping four miles out and four miles in along a turnpike-road, that his hapless spirits discoursed

> "Of fate, free-will, foreknowledge absolute,
> And found no end, in wandering mazes lost."

Seriously, if we wish rural walks to do our children any good, we must give them a love for rural sights, an object in every walk; we must teach them - and we can teach them - to find wonder in every insect, sublimity in every hedgerow, the records of past worlds in every pebble, and boundless fertility upon the barren shore; and so, by teaching them to make full use of that limited sphere in which they now are, make them faithful in a few things, that they may be fit hereafter to be rulers over much.

I may seem to exaggerate the advantages of such studies; but the question after all is one of experience: and I have had experience enough and to spare that what I say is true. I have seen the young man of fierce passions, and uncontrollable daring, expend healthily that energy which threatened daily to plunge him into recklessness, if not into sin, upon hunting out and collecting, through rock and bog, snow and tempest, every bird and egg of the neighbouring forest. I have seen the cultivated man, craving for travel and for success in life, pent up in the drudgery of London work, and yet keeping his spirit calm, and perhaps his morals all the more righteous, by spending over his microscope evenings which would too probably have gradually been wasted at the theatre. I have seen the young London beauty, amid all the excitement and temptation of luxury and flattery, with her heart pure and her mind occupied in a boudoir full of shells and fossils, flowers and sea-weeds; keeping herself unspotted from the world, by considering the lilies of the field, how they grow. And therefore it is that I hail with thankfulness every fresh book of Natural History, as a fresh boon to the young, a fresh help to those who have to educate them.

The greatest difficulty in the way of beginners is (as in most things) how "to learn the art of learning." They go out, search, find less than they expected, and give the subject up in disappointment. It is good to begin, therefore, if possible, by playing the part of "jackal" to some practised naturalist, who will show the tyro where to look, what to look for, and, moreover, what it is that he has found; often no easy matter to discover. Forty years ago, during an autumn's work of dead-leaf-searching in the Devon woods for poor old Dr. Turton, while he was writing his book on British land-shells, the present writer learnt more of the art of observing than he would have learnt in three years' desultory hunting on his own account; and he has often regretted that no naturalist has established shore-lectures at some watering-place, like those up hill and down dale field-lectures which, in pleasant bygone Cambridge days, Professor Sedgwick used to give to young geologists, and Professor Henslow to young botanists.

In the meanwhile, to show you something of what may be seen by those who care to see, let me take you, in imagination, to a shore where I was once at home, and for whose richness I can vouch, and choose our season and our day to start forth, on some glorious September or October morning, to see what last night's equinoctial gale has swept from the populous shallows of Torbay, and cast up, high and dry, on Paignton sands.

Torbay is a place which should be as much endeared to the naturalist as to the patriot and to the artist. We cannot gaze on its blue ring of water, and the great limestone bluffs which bound it to the north and south, without a glow passing through our hearts, as we remember the terrible and glorious pageant which passed by in the glorious July days of 1588, when the Spanish Armada ventured slowly past Berry Head, with Elizabeth's gallant pack of Devon captains (for the London fleet had not yet joined) following fast in its wake, and dashing into the midst of the vast line, undismayed by size and numbers, while their kin and friends stood watching and praying on the cliffs, spectators of Britain's Salamis. The white line of houses, too, on the other side of the bay, is Brixham, famed as the landing-place of William of Orange; the stone on the pier-head, which marks his first footsteps on British ground, is sacred in the eyes of all true English Whigs; and close by stands the castle of the settler of Newfoundland, Sir Humphrey Gilbert, Raleigh's half-brother, most learned of all Elizabeth's admirals in life, most pious and heroic in death. And as for scenery, though it can boast of neither mountain peak nor dark fiord, and would seem tame enough in the eyes of a western Scot or Irishman, yet Torbay surely has a soft beauty of its own.

The rounded hills slope gently to the sea, spotted with squares of emerald grass, and rich red fallow fields, and parks full of stately timber trees. Long lines of tall elms run down to the very water's edge, their boughs unwarped by any blast; here and there apple orchards are bending under their loads of fruit, and narrow strips of water-meadow line the glens, where the red cattle are already lounging in richest pastures, within ten yards of the rocky pebble beach. The shore is silent now, the tide far out: but six hours hence it will be hurling columns of rosy foam high into the sunlight, and sprinkling passengers, and cattle, and trim gardens which hardly know what frost and snow may be, but see the flowers of autumn meet the flowers of spring, and the old year linger smilingly to twine a garland for the new.

No wonder that such a spot as Torquay, with its delicious Italian climate, and endless variety of rich woodland, flowery lawn, fantastic rock-cavern, and broad bright tide-sand, sheltered from every wind of heaven except the soft south-east, should have become a favourite haunt, not only for invalids, but for naturalists.

Indeed, it may well claim the honour of being the original home of marine zoology and botany in England, as the Firth of Forth, under the auspices of Sir J. G. Dalyell, has been for Scotland. For here worked Montagu, Turton, and Mrs. Griffith, to whose extraordinary powers of research English marine botany almost owes its existence, and who survived to an age long beyond the natural term of man, to see, in her cheerful and honoured old age, that knowledge become popular and general which she pursued for many a year unassisted and alone. Here, too, the scientific succession is still maintained by Mr. Pengelly and Mr. Gosse, the latter of whom by his delightful and, happily, well-known books has done more for the study of marine zoology than any other living man. Torbay, moreover, from the variety of its rocks, aspects, and sea-floors, where limestones alternate with traps, and traps with slates, while at the valley-mouth the soft sandstones and hard conglomerates of the new red series slope down into the tepid and shallow waves, affords an abundance and variety of animal and vegetable life, unequalled, perhaps, in any other part of Great Britain. It cannot boast, certainly, of those strange deep-sea forms which Messrs.

Alder, Goodsir, and Laskey dredge among the lochs of the western Highlands, and the sub-marine mountain glens of the Zetland sea; but it has its own varieties, its own ever-fresh novelties: and in spite of all the research which has been lavished on its shores, a naturalist cannot, I suspect, work there for a winter without discovering forms new to science, or meeting with curiosities which have escaped all observers, since the lynx eye of Montagu espied them full fifty years ago.

Follow us, then, reader, in imagination, out of the gay watering-place, with its London shops and London equipages, along the broad road beneath the sunny limestone cliff, tufted with golden furze; past the huge oaks and green slopes of Tor Abbey; and past the fantastic rocks of Livermead, scooped by the waves into a labyrinth of double and triple caves, like Hindoo temples, upborne on pillars banded with yellow and white and red, a week's study, in form and colour and chiaro-oscuro, for any artist; and a mile or so further along a pleasant road, with land-locked glimpses of the bay, to the broad sheet of sand which lies between the village of Paignton and the sea - sands trodden a hundred times by Montagu and Turton, perhaps, by Dillwyn and Gaertner, and many another pioneer of science. And once there, before we look at anything else, come down straight to the sea marge; for yonder lies, just left by the retiring tide, a mass of life such as you will seldom see again.

It is somewhat ugly, perhaps, at first sight; for ankle-deep are spread, for some ten yards long by five broad, huge dirty bivalve shells, as large as the hand, each with its loathly grey and black siphons hanging out, a confused mass of slimy death. Let us walk on to some cleaner heap, and leave these, the great Lutraria Elliptica, which have been lying buried by thousands in the sandy mud, each with the point of its long siphon above the surface, sucking in and driving out again the salt water on which it feeds, till last night's ground-swell shifted the sea-bottom, and drove them up hither to perish helpless, but not useless, on the beach.

See, close by is another shell bed, quite as large, but comely enough to please any eye. What a variety of forms and colours are there, amid the purple and olive wreaths of wrack, and bladder-weed, and tangle (ore-weed, as they call it in the south), and the delicate green ribbons of the Zostera (the only English flowering plant which grows beneath the sea). What are they all? What are the long white razors? What are the delicate green-grey scimitars?

What are the tapering brown spires? What the tufts of delicate yellow plants like squirrels' tails, and lobsters' horns, and tamarisks, and fir-trees, and all other finely cut animal and vegetable forms? What are the groups of grey bladders, with something like a little bud at the tip? What are the hundreds of little pink-striped pears? What those tiny babies' heads, covered with grey prickles instead of hair? The great red star-fish, which Ulster children call "the bad man's hands;" and the great whelks, which the youth of Musselburgh know as roaring buckies, these we have seen before; but what, oh what, are the red capsicums? -

Yes, what are the red capsicums? and why are they poking, snapping, starting, crawling, tumbling wildly over each other, rattling about the huge mahogany cockles, as big as a child's two fists, out of which they are protruded? Mark them well, for you will perhaps never see them again. They are a Mediterranean species, or rather three species, left behind upon these extreme south-western coasts, probably at the vanishing of that warmer ancient epoch, which clothed the Lizard Point with the Cornish heath, and the Killarney mountains with Spanish saxifrages, and other relics of a flora whose home is now the Iberian peninsula and the sunny cliffs of the Riviera. Rare on every other shore, even in the west, it abounds in Torbay at certain, or rather uncertain, times, to so prodigious an amount, that the dredge, after five minutes' scrape, will sometimes come up choked full of this great cockle only. You will see hundreds of them in every cove for miles this day; a seeming waste of life, which would be awful, in our eyes, were not the Divine Ruler, as His custom is, making this destruction the means of fresh creation, by burying them in the sands, as soon as washed on shore, to fertilize the strata of some future world. It is but a shell-fish truly; but the great Cuvier thought it remarkable enough to devote to its anatomy elaborate descriptions and drawings, which have done more perhaps than any others to illustrate the curious economy of the whole class of bivalve, or double-shelled, mollusca. (Plate II. Fig. 3.)

That red capsicum is the foot of the animal contained in the cockleshell. By its aid it crawls, leaps, and burrows in the sand, where it lies drinking in the salt water through one of its siphons, and discharging it again through the other. Put the shell into a rock pool, or a basin of water, and you will see the siphons clearly. The valves gape apart some three-quarters of an inch. The semi-pellucid orange "mantle" fills the intermediate space. Through that mantle, at the end from which the foot curves, the siphons protrude; two thick short tubes joined side by side, their lips fringed with pearly cirri, or fringes; and very beautiful they are. The larger is always open, taking in the water, which is at once the animal's food and air, and which, flowing over the delicate inner surface of the mantle, at once oxygenates its blood, and fills its stomach with minute particles of decayed organized matter. The smaller is shut. Wait a minute, and it will open suddenly and discharge a jet of clear water, which has been robbed, I suppose, of its oxygen and its organic matter. But, I suppose, your eyes will be rather attracted by that same scarlet and orange foot, which is being drawn in and thrust out to a length of nearly four inches, striking with its point against any opposing object, and sending the whole shell backwards with a jerk. The point, you see, is sharp and tongue-like; only flattened, not horizontally, like a tongue, but perpendicularly, so as to form, as it was intended, a perfect sand-plough, by which the animal can move at will, either above or below the surface of the sand. [2]

But for colour and shape, to what shall we compare it? To polished cornelian, says Mr. Gosse. I say, to one of the great red capsicums which hang drying in every Covent-garden seedsman's window. Yet is either simile better than the guess of a certain lady, who, entering a room wherein a couple of Cardium tuberculatum were waltzing about a plate, exclaimed, "Oh dear! I always heard that my pretty red coral came out of a fish, and here it is all alive!"

"C. tuberculatum," says Mr. Gosse (who described it from specimens which I sent him in 1854), "is far the finest species. The valves are more globose and of a warmer colour; those that I have seen are even more spinous." Such may have been the case in those I sent: but it has occurred to me now and then to dredge specimens of C. aculeatum, which had escaped that rolling on the sand fatal in old age to its delicate spines, and which equalled in colour, size, and perfectness the noble one figured in poor dear old Dr. Turton's "British Bivalves." Besides, aculeatum is a far thinner and more delicate shell. And a third species, C. echinatum, with curves more graceful and continuous, is to be found now and then with the two former. In it, each point, instead of degenerating into a knot, as in tuberculatum, or developing from delicate flat briar-prickles into long straight thorns, as in aculeatum, is close-set to its fellow, and curved at the point transversely to the shell, the whole being thus horrid with hundreds of strong tenterhooks, making his castle impregnable to the raveners of the deep. For we can hardly doubt that these prickles are meant as weapons of defence, without which so savoury a morsel as the mollusc within (cooked and eaten largely on some parts of our south coast) would be a staple article of food for sea-beasts of prey. And it is noteworthy, first, that the defensive thorns which are permanent on the two thinner species, aculeatum and echinatum, disappear altogether on the thicker one, tuberculatum, as old age gives him a solid and heavy globose shell; and next, that he too, while young and tender, and liable therefore to be bored through by whelks and such murderous univalves, does actually possess the same briar- prickles, which his thinner cousins keep throughout life. Nevertheless, prickles, in all three species, are, as far as we can see, useless in Torbay, where no wolf-fish (Anarrhichas lupus) or other owner of shell-crushing jaws wanders, terrible to lobster and to cockle. Originally intended, as we suppose, to face the strong- toothed monsters of the Mediterranean, these foreigners have wandered northward to shores where their armour is not now needed; and yet centuries of idleness and security have not been able to persuade them to lay it by. This - if my explanation is the right one - is but one more case among hundreds in which peculiarities, useful doubtless to their original possessors, remain, though now useless, in their descendants. Just so does the tame ram inherit the now superfluous horns of his primeval wild ancestors, though he fights now - if he fights at all - not with his horns, but with his forehead.

Enough of Cardium tuberculatum. Now for the other animals of the heap; and first, for those long white razors. They, as well as the grey scimitars, are Solens, Razor-fish (Solen siliqua and S. ensis), burrowers in the sand by that foot which protrudes from one end, nimble in escaping from the Torquay boys, whom you will see boring for them with a long iron screw, on the sands at low tide. They are very good to eat, these razor-fish; at least, for those who so think them; and abound in millions upon all our sandy shores. [3]

Now for the tapering brown spires. They are Turritellae, snail- like animals (though the form of the shell is different), who crawl and browse by thousands on the beds of Zostera, or grass wrack, which you see thrown about on the beach, and which grows naturally in two or three fathoms water. Stay: here is one which is "more than itself." On its back is mounted a cluster of barnacles (Balanus Porcatus), of the same family as those which stud the tide-rocks in millions, scratching the legs of hapless bathers. Of them, I will speak presently; for I may have a still more curious member of the family to show you. But meanwhile, look at the mouth of the shell; a long grey worm protrudes from it, which is not the rightful inhabitant. He is dead long since, and his place has been occupied by one Sipunculus Bernhardi; a wight of low degree, who connects "radiate" with annulate forms - in plain English, sea- cucumbers (of which we shall see some soon) with sea-worms. But however low in the scale of comparative anatomy, he has wit enough to take care of himself; mean ugly little worm as he seems. For finding the mouth of the Turritella too big for him, he has plastered it up with sand and mud (Heaven alone knows how), just as a wry-neck plasters up a hole in an apple-tree when she intends to build therein, and has left only a round hole, out of which he can poke his proboscis. A curious thing is this proboscis, when seen through the magnifier. You perceive a ring of tentacles round the mouth, for picking up I know not what; and you will perceive, too, if you watch it, that when he draws it in, he turns mouth, tentacles and all, inwards, and so down into his stomach, just as if you were to turn the finger of a glove inward from the tip till it passed into the hand; and so performs, every time he eats, the clown's as yet ideal feat, of jumping down his own throat. [4]

So much have we seen on one little shell. But there is more to see close to it. Those yellow plants which I likened to squirrels' tails and lobsters' horns, and what not, are zoophytes of different kinds. Here is Sertularia argentea (true squirrel's tail); here, S. filicula, as delicate as tangled threads of glass; here, abietina; here, rosacea. The lobsters' horns are Antennaria antennina; and mingled with them are Plumulariae, always to be distinguished from Sertulariae by polypes growing on one side of the branch, and not on both. Here is falcata, with its roots twisted round a sea-weed. Here is cristata, on the same weed; and here is a piece of the beautiful myriophyllum, which has been battered in its long journey out of the deep water about the ore rock. For all these you must consult Johnson's "Zoophytes," and for a dozen smaller species, which you would probably find tangled among them, or parasitic on the sea-weed. Here are Flustrae, or sea-mats. This, which smells very like Verbena, is Flustra coriacea (Pl. I. Fig. 2). That scurf on the frond of ore-weed is F. lineata (Pl. Fig. 1). The glass bells twined about this Sertularia are Campanularia syringa (Pl. I. Fig. 9); and here is a tiny plant of Cellularia ciliata (Pl. I. Fig. 8). Look at it through the field-glass; for it is truly wonderful. Each polype cell is edged with whip-like spines, and on the back of some of them is - what is it, but a live vulture's head, snapping and snapping - what for?

Nay, reader, I am here to show you what can be seen: but as for telling you what can be known, much more what cannot, I decline; and refer you to Johnson's "Zoophytes," wherein you will find that several species of polypes carry these same birds' heads: but whether they be parts of the polype, and of what use they are, no man living knoweth.

Next, what are the striped pears? They are sea-anemones, and of a species only lately well known, Sagartia viduata, the snake-locked anemone (Pl. V. Fig. 3[5]). They have been washed off the loose stones to which they usually adhere by the pitiless roll of the ground-swell; however, they are not so far gone, but that if you take one of them home, and put it in a jar of water, it will expand into a delicate compound flower, which can neither be described nor painted, of long pellucid tentacles, hanging like a thin bluish cloud over a disk of mottled brown and grey.

Here, adhering to this large whelk, is another, but far larger and coarser. It is Sagartia parasitica, one of our largest British species; and most singular in this, that it is almost always (in Torbay, at least,) found adhering to a whelk: but never to a live one; and for this reason. The live whelk (as you may see for yourself when the tide is out) burrows in the sand in chase of hapless bivalve shells, whom he bores through with his sharp tongue (always, cunning fellow, close to the hinge, where the fish is), and then sucks out their life. Now, if the anemone stuck to him, it would be carried under the sand daily, to its own disgust. It prefers, therefore, the dead whelk, inhabited by a soldier crab, Pagurus Bernhardi (Pl. II. Fig. 2), of which you may find a dozen anywhere as the tide goes out; and travels about at the crab's expense, sharing with him the offal which is his food. Note, moreover, that the soldier crab is the most hasty and blundering of marine animals, as active as a monkey, and as subject to panics as a horse; wherefore the poor anemone on his back must have a hard life of it; being knocked about against rocks and shells, without warning, from morn to night and night to morn. Against which danger, kind Nature, ever MAXIMA IN MINIMIS, has provided by fitting him with a stout leather coat, which she has given, I believe, to no other of his family.

Next, for the babies' heads, covered with prickles, instead of hair. They are sea-urchins, Amphidotus cordatus, which burrow by thousands in the sand. These are of that Spatangoid form, which you will often find fossil in the chalk, and which shepherd boys call snakes' heads. We shall soon find another sort, an Echinus, and have time to talk over these most strange (in my eyes) of all living animals.

There are a hundred more things to be talked of here: but we must defer the examination of them till our return; for it wants an hour yet of the dead low spring-tide; and ere we go home, we will spend a few minutes at least on the rocks at Livermead, where awaits us a strong-backed quarryman, with a strong-backed crowbar, as is to be hoped (for he snapped one right across there yesterday, falling miserably on his back into a pool thereby), and we will verify Mr. Gosse's observation, that -

> "When once we have begun to look with curiosity on the strange things that ordinary people pass over without notice, our wonder is continually excited by the variety of phase, and often by the uncouthness of form, under which some of the meaner creatures are presented to us. And this is very specially the case with the inhabitants of the sea. We can scarcely poke or pry for an hour among the rocks, at low-water mark, or walk, with an observant downcast eye, along the beach after a gale, without finding some oddly-fashioned, suspicious-looking being, unlike any form of life that we have seen before. The dark concealed interior of the sea becomes thus invested with a fresh mystery; its vast recesses appear to be stored with all imaginable forms; and we are tempted to think there must be multitudes of living creatures whose very figure and structure have never yet been suspected.
>
> "'O sea! old sea! who yet knows half
> Of thy wonders or thy pride!'"
> GOSSE'S AQUARIUM, pp. 226, 227.

These words have more than fulfilled themselves since they were written. Those Deep-Sea dredgings, of which a detailed account will be found in Dr. Wyville Thomson's new and most beautiful book, "The Depths of the Sea," have disclosed, of late years, wonders of the deep even more strange and more multitudinous than the wonders of the shore. The time is past when we thought ourselves bound to believe, with Professor Edward Forbes, that only some hundred fathoms down, the inhabitants of the sea-bottom "become more and more modified, and fewer and fewer, indicating our approach towards an abyss where life is either extinguished, or exhibits but a few sparks to mark it's lingering presence."

Neither now need we indulge in another theory which had a certain grandeur in it, and was not so absurd as it looks at first sight, - namely, that, as Dr. Wyville Thomson puts it, picturesquely enough, "in going down the sea water became, under the pressure, gradually heavier and heavier, and that all the loose things floated at different levels, according to their specific weight, - skeletons of men, anchors and shot and cannon, and last of all the broad gold pieces lost in the wreck of many a galleon off the Spanish Main; the whole forming a kind of 'false bottom' to the ocean, beneath which there lay all the depth of clear still water, which was heavier than molten gold."

The facts are; first that water, being all but incompressible, is hardly any heavier, and just as liquid, at the greatest depth, than at the surface; and that therefore animals can move as freely in it in deep as in shallow water; and next, that as the fluids inside the body of a sea animal must be at the same pressure as that of the water outside it, the two pressures must balance each other; and the body, instead of being crushed in, may be unconscious that it is living under a weight of two or three miles of water. But so it is; as we gather our curiosities at low-tide mark, or haul the dredge a mile or two out at sea, we may allow our fancy to range freely out to the westward, and down over the subaqueous cliffs of the hundred-fathom line, which mark the old shore of the British Isles, or rather of a time when Britain and Ireland were part of the continent, through water a mile, and two, and three miles deep, into total darkness, and icy cold, and a pressure which, in the open air, would crush any known living creature to a jelly; and be certain that we shall find the ocean-floor teeming everywhere with multitudinous life, some of it strangely like, some strangely unlike, the creatures which we see along the shore.

Some strangely like. You may find, for instance, among the sea-weed, here and there, a little black sea-spider, a Nymphon, who has this peculiarity, that possessing no body at all to speak of, he carries his needful stomach in long branches, packed inside his legs. The specimens which you will find will probably be half an inch across the legs. An almost exactly similar Nymphon has been dredged from the depths of the Arctic and Antarctic oceans, nearly two feet across.

You may find also a quaint little shrimp, CAPRELLA, clinging by its hind claws to sea-weed, and waving its gaunt grotesque body to and fro, while it makes mesmeric passes with its large fore claws, - one of the most ridiculous of Nature's many ridiculous forms. Those which you will find will be some quarter of an inch in length; but in the cold area of the North Atlantic, their cousins, it is now found, are nearly three inches long, and perch in like manner, not on sea-weeds, for there are none so deep, but on branching sponges.

These are but two instances out of many of forms which were supposed to be peculiar to shallow shores repeating themselves at vast depths: thus forcing on us strange questions about changes in the distribution and depth of the ancient seas; and forcing us, also, to reconsider the old rules by which rocks were distinguished as deep-sea or shallow-sea deposits according to the fossils found in them.

As for the new forms, and even more important than them, the ancient forms, supposed to have been long extinct, and only known as fossils, till they were lately rediscovered alive in the nether darkness, - for them you must consult Dr. Wyville Thomson's book, and the notices of the "Challenger's" dredgings which appear from time to time in the columns of "Nature;" for want of space forbids my speaking of them here.

But if you have no time to read "The Depths of the Sea," go at least to the British Museum, or if you be a northern man, to the admirable public museum at Liverpool; ask to be shown the deep-sea forms; and there feast your curiosity and your sense of beauty for an hour. Look at the Crinoids, or stalked star-fishes, the "Lilies of living stone," which swarmed in the ancient seas, in vast variety, and in such numbers that whole beds of limestone are composed of their disjointed fragments; but which have vanished out of our modern seas, we know not why, till, a few years since, almost the only known living species was the exquisite and rare Pentacrinus asteria, from deep water off the Windward Isles of the West Indies.

Of this you will see a specimen or two both at Liverpool and in the British Museum; and near them, probably, specimens of the new-old Crinoids, discovered of late years by Professor Sars, Mr. Gwyn Jeffreys, Dr. Carpenter, Dr. Wyville Thomson, and the other deep- sea disciples of the mythic Glaucus, the fisherman, who, enamoured of the wonders of the sea, plunged into the blue abyss once and for all, and became himself "the blue old man of the sea."

Next look at the corals, and Gorgonias, and all the sea-fern tribe of branching polypidoms, and last, but not least, at the glass sponges; first at the Euplectella, or Venus's flower-basket, which lives embedded in the mud of the seas of the Philippines, supported by a glass frill "standing up round it like an Elizabethan ruff."

Twenty years ago there was but one specimen in Europe: now you may buy one for a pound in any curiosity shop. I advise you to do so, and to keep - as I have seen done - under a glass case, as a delight to your eyes, one of the most exquisite, both for form and texture, of natural objects.

Then look at the Hyalonemas, or glass-rope ocean floor by a twisted wisp of strong flexible flint needles, somewhat on the principle of a screw-pile. So strange and complicated is their structure, that naturalists for a long while could literally make neither head nor tail of them, as long as they had only Japanese specimens to study, some of which the Japanese dealers had, of malice prepense, stuck upside down into Pholas-borings in stones. Which was top and which bottom; which the thing itself, and which special parasites growing on it; whether it was a sponge, or a zoophyte, or something else; at one time even whether it was natural, or artificial and a make- up, - could not be settled, even till a year or two since. But the discovery of the same, or a similar, species in abundance from the Butt of the Lows down to Setubal on the Portuguese coast, where the deep-water shark fishers call it "sea-whip," has given our savants specimens enough to make up their minds - that they really know little or nothing about it, and probably will never know.

And do not forget, lastly, to ask, whether at Liverpool or at the British Museum, for the Holtenias and their congeners, - hollow sponges built up of glassy spicules, and rooted in the mud by glass hairs, in some cases between two and three feet long, as flexible and graceful as tresses of snow-white silk.

Look at these, and a hundred kindred forms, and then see how nature is not only "maxima in minimis" - greatest in her least, but often "pulcherrima in abditis" - fairest in her most hidden works; and how the Creative Spirit has lavished, as it were, unspeakable artistic skill on lowly-organized creature, never till now beheld by man, and buried, not only in foul mud, but in their own unsightly heap of living jelly.

But so it was from the beginning; - and this planet was not made for man alone. Countless ages before we appeared on earth the depths of the old chalk-ocean teemed with forms as beautiful and perfect as those, their lineal descendants, which the dredge now brings up from the Atlantic sea-floor; and if there were - as my reason tells me that there must have been - final moral causes for their existence, the only ones which we have a right to imagine are these - that all, down to the lowest Rhizopod, might delight themselves, however dimly, in existing; and that the Lord might delight Himself in them.

Thus, much - alas! how little - about the wonders of the deep. We, who are no deep-sea dredgers, must return humbly to the wonders of the shore. And first, as after descending the gap in the sea-wall we walk along the ribbed floor of hard yellow sand, let me ask you to give a sharp look-out for a round grey disc, about as big as a penny-piece, peeping out on the surface. No; that is not it, that little lump: open it, and you will find within one of the common little Venus gallina. - The closet collectors have given it some new name now, and no thanks to them: they are always changing the names, instead of studying the live animals where Nature has put them, in which case they would have no time for word-inventing. Nay, I verify suspect that the names grow, like other things; at least, they get longer and longer and more jaw-breaking every year. The little bivalve, however, finding itself left by the tide, has wisely shut up its siphons, and, by means of its foot and its edges, buried itself in a comfortable bath of cool wet sand, till the sea shall come back, and make it safe to crawl and lounge about on the surface, smoking the sea-water instead of tobacco. Neither is that depression what we seek. Touch it, and out poke a pair of astonished and inquiring horns: it is a long-armed crab, who saw us coming, and wisely shovelled himself into the sand by means of his nether-end. Corystes Cassivelaunus is his name, which he is said to have acquired from the marks on his back, which are somewhat like a human face. "Those long antennae," says my friend, Mr. Lloyd [6] - I have not verified the fact, but believe it, as he knows a great deal about crabs, and I know next to nothing - "form a tube through which a current of water passes into the crab's gills, free from the surrounding sand." Moreover, it is only the male who has those strangely long fore-arms and claws; the female contenting herself with limbs of a more moderate length. Neither is that, though it might be, the hole down which what we seek has vanished: but that burrow contains one of the long white razors which you saw cast on shore at Paignton. The boys close by are boring for them with iron rods armed with a screw, and taking them in to sell in Torquay market, as excellent food. But there is one, at last - a grey disc pouting up through the sand. Touch it, and it is gone down, quick as light. We must dig it out, and carefully, for it is a delicate monster. At last, after ten minutes' careful work, we have brought up, from a foot depth or more - what? A thick, dirty, slimy worm, without head or tail, form or colour. A slug has more artistic beauty about him. Be it so. At home in the aquarium (where, alas! he will live but for a day or two, under the new irritation of light) he will make a very different figure. That is one of the rarest of British sea- animals, Peachia hastata (Pl. XII. Fig. 1), which differs from most other British Actiniae in this, that instead of having like them a walking disc, it has a free open lower end, with which (I know not how) it buries itself upright in the sand, with its mouth just above the surface. The figure on the left of the plate represents a curious cluster of papillae which project from one side of the mouth, and are the opening of the oviduct. But his value consists, not merely in his beauty (though that, really, is not small), but in his belonging to what the long word-makers call an "interosculant" group, - a party of genera and species which connect families scientifically far apart, filling up a fresh link in the great chain, or rather the great network, of zoological classification. For here we have a simple, and, as it were, crude form; of which, if we dared to indulge in reveries, we might say that the Creative Mind realized it before either Actiniae or Holothurians, and then went on to perfect the idea contained in it in two different directions; dividing it into two different families, and making on its model, by adding new organs, and taking away old ones, in one direction the whole family of Actiniae (sea-anemones), and in a quite opposite one the Holothuriae, those strange sea-cucumbers, with their mouth-fringe of feathery gills, of which you shall see some anon. Thus there has been, in the Creative Mind, as it gave life to new species, a development of the idea on which older species were created, in order - we may fancy - that every mesh of the great net might gradually be supplied, and there should be no gaps in the perfect variety of Nature's forms. This development is one which we must believe to be at least possible, if we allow that a Mind presides over the universe, and not a mere brute necessity, a Law (absurd misnomer) without a Lawgiver; and to it (strangely enough coinciding here and there with the Platonic doctrine of Eternal Ideas existing in the Divine Mind) all fresh inductive discovery seems to point more and more.

Let me speak freely a few words on this important matter. Geology has disproved the old popular belief that the universe was brought into being as it now exists by a single fiat. We know that the work has been gradual; that the earth

"In tracts of fluent heat began,
The seeming prey of cyclic storms,
The home of seeming random forms,
Till, at the last, arose the man."

And we know, also, that these forms, "seeming random" as they are, have appeared according to a law which, as far as we can judge, has been on the whole one of progress, - lower animals (though we cannot yet say, the lowest) appearing first, and man, the highest mammal, "the roof and crown of things," one of the latest in the series. We have no more right, let it be observed, to say that man, the highest, appeared last, than that the lowest appeared first. It was probably so, in both cases; but there is as yet no positive proof of either; and as we know that species of animals lower than those which already existed appeared again and again during the various eras, so it is quite possible that they may be appearing now, and may appear hereafter: and that for every extinct Dodo or Moa, a new species may be created, to keep up the equilibrium of the whole. This is but a surmise: but it may be wise, perhaps, just now, to confess boldly, even to insist on, its possibility, lest any should fancy, from our unwillingness to allow it, that there would be ought in it, if proved, contrary to sound religion.

I am, I must honestly confess, more and more unable to perceive anything which an orthodox Christian may not hold, in those physical theories of "evolution," which are gaining more and more the assent of our best zoologists and botanists. All that they ask us to believe is, that "species" and "families," and indeed the whole of organic nature, have gone through, and may still be going through, some such development from a lowest germ, as we know that every living individual, from the lowest zoophyte to man himself, does actually go through. They apply to the whole of the living world, past, present, and future, the law which is undeniably at work on each individual of it. They may be wrong, or they may be right: but what is there in such a conception contrary to any doctrine - at least of the Church of England? To say that this cannot be true; that species cannot vary, because God, at the beginning, created each thing "according to its kind," is really to beg the question; which is - Does the idea of "kind" include variability or not? and if so, how much variability? Now, "kind," or "species," as we call it, is defined nowhere in the Bible. What right have we to read our own definition into the word? - and that against the certain fact, that some "kinds" do vary, and that widely, - mankind, for instance, and the animals and plants which he domesticates. Surely that latter fact should be significant, to those who believe, as I do, that man was created in the likeness of God. For if man has the power, not only of making plants and animals vary, but of developing them into forms of higher beauty and usefulness than their wild ancestors possessed, why should not the God in whose image he is made possess the same power? If the old theological rule be true - "There is nothing in man which was not first in God" (sin, of course, excluded) - then why should not this imperfect creative faculty in man be the very guarantee that God possesses it in perfection?

Such at least is the conclusion of one who, studying certain families of plants, which indulge in the most fantastic varieties of shape and size, and yet through all their vagaries retain - as do the Palms, the Orchids, the Euphorbiaceae - one organ, or form of organs, peculiar and highly specialized, yet constant throughout the whole of each family, has been driven to the belief that each of these three families, at least, has "sported off" from one common ancestor - one archetypal Palm, one archetypal Orchid, one archetypal Euphorbia, simple, it may be, in itself, but endowed with infinite possibilities of new and complex beauty, to be developed, not in it, but in its descendants. He has asked himself, sitting alone amid the boundless wealth of tropic forests, whether even then and there the great God might not be creating round him, slowly but surely, new forms of beauty? If he chose to do it, could He not do it? That man found himself none the worse Christian for the thought. He has said - and must be allowed to say again, for he sees no reason to alter his words - in speaking of the wonderful variety of forms in the Euphorbiaceae, from the weedy English Euphorbias, the Dog's Mercuries, and the Box, to the prickly-stemmed Scarlet Euphorbia of Madagascar, the succulent Cactus-like Euphorbias of the Canaries and elsewhere; the Gale-like Phyllanthus; the many-formed Crotons; the Hemp-like Maniocs, Physic-nuts, Castor-oils, the scarlet Poinsettia, the little pink and yellow Dalechampia, the poisonous Manchineel, and the gigantic Hura, or sandbox tree, of the West Indies, - all so different in shape and size, yet all alike in their most peculiar and complex fructification, and in their acrid milky juice,- "What if all these forms are the descendants of one original form? Would that be one whit the more wonderful than the theory that they were, each and all, with the minute, and often imaginary, shades of difference between certain cognate species among them, created separately and at once? But if it be so - which I cannot allow - what would the theologian have to say, save that God's works are even more wonderful than he always believed them to be? As for the theory being impossible - that is to be decided by men of science, on strict experimental grounds. As for us theologians, who are we, that we should limit, ... priori, the power of God? 'Is anything too hard for the Lord?' asked the prophet of old; and we have a right to ask it as long as the world shall last. If it be said that 'natural selection,' or, as Mr. Herbert Spencer better defines it, the 'survival of the fittest,' is too simple a cause to produce such fantastic variety - that, again, is a question to be settled exclusively by men of science, on their own grounds. We, meanwhile, always knew that God works by very simple, or seemingly simple, means; that the universe, as far as we could discern it, was one organization of the most simple means. It was wonderful - or should have been - in our eyes, that a shower of rain should make the grass grow, and that the grass should become flesh, and the flesh food for the thinking brain of man. It was - or ought to have been - more wonderful yet to us that a child should resemble its parents, or even a butterfly resemble, if not always, still usually, its parents likewise. Ought God to appear less or more august in our eyes if we discover that the means are even simpler than we supposed? We held Him to be Almighty and All-wise. Are we to reverence Him less or more if we find Him to be so much mightier, so much wiser, than we dreamed, that He can not only make all things, but - the very perfection of creative power - MAKE ALL THINGS MAKE THEMSELVES? We believed that His care was over all His works; that His providence worked perpetually over the universe. We were taught - some of us at least - by Holy Scripture, that without Him not a sparrow fell to the ground, and that the very hairs of our head were all numbered; that the whole history of the universe was made up, in fact, of an infinite network of special providences. If, then, that should be true which a great naturalist writes, 'It may be metaphorically said that natural selection is daily and hourly scrutinizing, throughout the world, every variation, even the slightest; rejecting that which is bad, preserving and adding up all that is good; silently and insensibly working, whenever and wherever opportunity offers, at the improvement of each organic being, in relation to its organic and inorganic conditions of life,' - if this, I say, were proved to be true, ought God's care and God's providence to seem less or more magnificent in our eyes? Of old it was said by Him without whom nothing is made - 'My Father worketh hitherto, and I work.' Shall we quarrel with physical science, if she gives us evidence that those words are true?"

And - understand it well - the grand passage I have just quoted need not be accused of substituting "natural selection for God." In any case natural selection would be only the means or law by which God works, as He does by other natural laws. We do not substitute gravitation for God, when we say that the planets are sustained in their orbits by the law of gravitation. The theory about natural selection may be untrue, or imperfect, as may the modern theories of the "evolution and progress" of organic forms: let the man of science decide that. But if true, the theories seem to me perfectly to agree with, and may be perfectly explained by, the simple old belief which the Bible sets before us, of a LIVING GOD: not a mere past will, such as the Koran sets forth, creating once and for all, and then leaving the universe, to use Goethe's simile, "to spin round his finger;" nor again, an "all-pervading spirit," words which are mere contradictory jargon, concealing, from those who utter them, blank Materialism: but One who works in all things which have obeyed Him to will and to do of His good pleasure, keeping His abysmal and self-perfect purpose, yet altering the methods by which that purpose is attained, from aeon to aeon, ay, from moment to moment, for ever various, yet for ever the same. This great and yet most blessed paradox of the Changeless God, who yet can say "It repenteth me," and "Behold, I work a new thing on the earth," is revealed no less by nature than by Scripture; the changeableness, not of caprice or imperfection, but of an Infinite Maker and "Poietes," drawing ever fresh forms out of the inexhaustible treasury of His primaeval Mind; and yet never throwing away a conception to which He has once given actual birth in time and space, (but to compare reverently small things and great) lovingly repeating it, re-applying it; producing the same effects by endlessly different methods; or so delicately modifying the method that, as by the turn of a hair, it shall produce endlessly diverse effects; looking back, as it were, ever and anon over the great work of all the ages, to retouch it, and fill up each chasm in the scheme, which for some good purpose had been left open in earlier worlds; or leaving some open (the forms, for instance, necessary to connect the bimana and the quadrumana) to be filled up perhaps hereafter when the world needs them; the handiwork, in short, of a living and loving Mind, perfect in His own eternity, but stooping to work in time and space, and there rejoicing Himself in the work of His own hands, and in His eternal Sabbaths ceasing in rest ineffable, that He may look on that which He hath made, and behold it is very good.

I speak, of course, under correction; for this conclusion is emphatically matter of induction, and must be verified or modified by ever-fresh facts: but I meet with many a Christian passage in scientific books, which seems to me to go, not too far, but rather not far enough, in asserting the God of the Bible, as Saint Paul says, "not to have left Himself without witness," in nature itself, that He is the God of grace. Why speak of the God of nature and the God of grace as two antithetical terms? The Bible never, in a single instance, makes the distinction; and surely, if God be (as He is) the Eternal and Unchangeable One, and if (as we all confess) the universe bears the impress of His signet, we have no right, in the present infantile state of science, to put arbitrary limits of our own to the revelation which He may have thought good to make of Himself in nature. Nay, rather, let us believe that, if our eyes were opened, we should fulfil the requirement of Genius, to "see the universal in the particular," by seeing God's whole likeness, His whole glory, reflected as in a mirror even in the meanest flower; and that nothing but the dulness of our own souls prevents them from seeing day and night in all things, however small or trivial to human eclecticism, the Lord Jesus Christ Himself fulfilling His own saying, "My Father worketh hitherto, and I work."

To me it seems (to sum up, in a few words, what I have tried to say) that such development and progress as have as yet been actually discovered in nature, bear every trace of having been produced by successive acts of thought and will in some personal mind; which, however boundlessly rich and powerful, is still the Archetype of the human mind; and therefore (for to this I confess I have been all along tending) probably capable, without violence to its properties, of becoming, like the human mind, incarnate.

But to descend from these perhaps too daring speculations, there is another, and more human, source of interest about the animal who is writhing feebly in the glass jar of salt water; for he is one of the many curiosities which have been added to our fauna by that humble hero Mr. Charles Peach, the self-taught naturalist, of whom, as we walk on toward the rocks, something should be said, or rather read; for Mr. Chambers, in an often-quoted passage from his Edinburgh Journal, which I must have the pleasure of quoting once again, has told the story better than we can tell it:-

> "But who is that little intelligent-looking man in a faded naval uniform, who is so invariably to be seen in a particular central seat in this section? That, gentle

reader, is perhaps one of the most interesting men who attend the British Association. He is only a private in the mounted guard (preventive service) at an obscure part of the Cornwall coast, with four shillings a day, and a wife and nine children, most of whose education he has himself to conduct. He never tastes the luxuries which are so common in the middle ranks of life, and even amongst a large portion of the working classes. He has to mend with his own hands every sort of thing that can break or wear in his house. Yet Mr. Peach is a votary of Natural History; not a student of the science in books, for he cannot afford books; but an investigator by sea and shore, a collector of Zoophytes and Echinodermata - strange creatures, many of which are as yet hardly known to man. These he collects, preserves, and describes; and every year does he come up to the British Association with a few novelties of this kind, accompanied by illustrative papers and drawings: thus, under circumstances the very opposite of those of such men as Lord Enniskillen, adding, in like manner, to the general stock of knowledge. On the present occasion he is unusually elated, for he has made the discovery of a Holothuria with twenty tentacula, a species of the Echinodermata which Professor Forbes, in his book on Star-Fishes, has said was never yet observed in the British seas. It may be of small moment to you, who, mayhap, know nothing of Holothurias: but it is a considerable thing to the Fauna of Britain, and a vast matter to a poor private of the Cornwall mounted guard. And accordingly he will go home in a few days, full of the glory of his exhibition, and strong anew by the kind notice taken of him by the masters of the science, to similar inquiries, difficult as it may be to prosecute them, under such a complication of duties, professional and domestic. Honest Peach! humble as is thy home, and simple thy bearing, thou art an honour even to this assemblage of nobles and doctors: nay, more, when we consider everything, thou art an honour to human nature itself; for where is the heroism like that of virtuous, intelligent, independent poverty? And such heroism is thine!" - CHAMBERS' EDIN. JOURN., Nov. 23, 1844.

Mr. Peach has been since rewarded in part for his long labours in the cause of science, by having been removed to a more lucrative post on the north coast of Scotland; the earnest, it is to be hoped, of still further promotion.

I mentioned just now Synapta; or, as Montagu called it, Chirodota: a much better name, and, I think, very uselessly changed; for Chirodota expresses the peculiarity of the beast, which consists in - start not, reader - twelve hands, like human hands, while Synapta expresses merely its power of clinging to the fingers, which it possesses in common with many other animals. It is, at least, a beast worth talking about; as for finding one, I fear that we have no chance of such good fortune.

Colonel Montagu found them here some forty years ago; and after him, Mr. Alder, in 1845. I found hundreds of them, but only once, in 1854 after a heavy south-eastern gale, washed up among the great Lutrariae in a cove near Goodrington; but all my dredging outside failed to procure a specimen - Mr. Alder, however, and Mr. Cocks (who find everything, and will at last certainly catch Midgard, the great sea-serpent, as Thor did, by baiting for him with a bull's head), have dredged them in great numbers; the former, at Helford in Cornwall, the latter on the west coast of Scotland. It seems, however, to be a southern monster, probably a remnant, like the great cockle, of the Mediterranean fauna; for Mr. MacAndrew finds them plentifully in Vigo Bay, and J. Moller in the Adriatic, off Trieste.

But what is it like? Conceive a very fat short earth-worm; not ringed, though, like the earth-worm, but smooth and glossy, dappled with darker spots, especially on one side, which may be the upper one. Put round its mouth twelve little arms, on each a hand with four ragged fingers, and on the back of the hand a stump of a thumb, and you have Synapta Digitata (Plates IV. and V., from my drawings of the live animal). These hands it puts down to its mouth, generally in alternate pairs, but how it obtains its food by them is yet a mystery, for its intestines are filled, like an earth-worm's, with the mud in which it lives, and from which it probably extracts (as does the earth-worm) all organic matters.

You will find it stick to your fingers by the whole skin, causing, if your hand be delicate, a tingling sensation; and if you examine the skin under the microscope, you will find the cause. The whole skin is studded with minute glass anchors, some hanging freely from the surface, but most imbedded in the skin. Each of these anchors is jointed at its root into one end of a curious cribriform plate, - in plain English, one pierced like a sieve, which lies under the skin, and reminds one of the similar plates in the skin of the White Cucumaria, which I will show you presently; and both of these we must regard as the first rudiments of an Echinoderm's outside skeleton, such as in the Sea-urchins covers the whole body of the animal. (See on Echinus Millaris, p. 89.) [7] Somewhat similar anchor-plates, from a Red Sea species, Synapta Vittata, may be seen in any collection of microscopic objects.

The animal, when caught, has a strange habit of self-destruction, contracting its skin at two or three different points, and writhing till it snaps itself into "junks," as the sailors would say, and then dies. My specimens, on breaking up, threw out from the wounded part long "ovarian filaments" (whatsoever those may be), similar to those thrown out by many of the Sagartian anemones, especially S. parasitica. Beyond this, I can tell you nothing about Synapta, and only ask you to consider its hands, as an instance of that fantastic play of Nature which repeats, in families widely different, organs of similar form, though perhaps of by no means similar use; nay, sometimes (as in those beautiful clear-wing hawk-moths which you, as they hover round the rhododendrons, mistake for bumble-bees) repeats the outward form of a whole animal, for no conceivable reason save her - shall we not say honestly His? - own good pleasure.

But here we are at the old bank of boulders, the ruins of an antique pier which the monks of Tor Abbey built for their convenience, while Torquay was but a knot of fishing huts within a lonely limestone cove. To get to it, though, we have passed many a hidden treasure; for every ledge of these flat New-red-sandstone rocks, if torn up with the crowbar, discloses in its cracks and crannies nests of strange forms which shun the light of day; beautiful Actiniae fill the tiny caverns with living flowers; great Pholades (Plate X. figs. 3, 4) bore by hundreds in the softer strata; and wherever a thin layer of muddy sand intervenes between two slabs, long Annelid worms of quaintest forms and colours have their horizontal burrows, among those of that curious and rare radiate animal, the Spoonworm, [8] an eyeless bag about an inch long, half bluish grey, half pink, with a strange scalloped and wrinkled proboscis of saffron colour, which serves, in some mysterious way, soft as it is, to collect food, and clear its dark passage through the rock.

See, at the extreme low-water mark, where the broad olive fronds of the Laminariae, like fan-palms, droop and wave gracefully in the retiring ripples, a great boulder which will serve our purpose.

Its upper side is a whole forest of sea-weeds, large and small; and that forest, if you examined it closely, as full of inhabitants as those of the Amazon or the Gambia. To "beat" that dense cover would be an endless task: but on the under side, where no sea- weeds grow, we shall find full in view enough to occupy us till the tide returns. For the slab, see, is such a one as sea-beasts love to haunt. Its weed-covered surface shows that the surge has not shifted it for years past. It lies on other boulders clear of sand and mud, so that there is no fear of dead sea-weed having lodged and decayed under it, destructive to animal life. We can see dark crannies and caves beneath; yet too narrow to allow the surge to wash in, and keep the surface clean. It will be a fine menagerie of Nereus, if we can but turn it.

Now the crowbar is well under it; heave, and with a will; and so, after five minutes' tugging, propping, slipping, and splashing, the boulder gradually tips over, and we rush greedily upon the spoil.

A muddy dripping surface it is, truly, full of cracks and hollows, uninviting enough at first sight: let us look it round leisurely, to see if there are not materials enough there for an hour's lecture.

The first object which strikes the eye is probably a group of milk- white slugs, from two to six inches long, cuddling snugly together (Plate IX. fig. 1). You try to pull them off, and find that they give you some trouble, such a firm hold have the delicate white sucking arms, which fringe each of their five edges. You see at the head nothing but a yellow dimple; for eating and breathing are suspended till the return of tide; but once settled in a jar of salt-water, each will protrude a large chocolate-coloured head, tipped with a ring of ten feathery gills, looking very much like a head of "curled kale," but of the loveliest white and primrose; in the centre whereof lies perdu a mouth with sturdy teeth - if indeed they, as well as the whole inside of the beast, have not been lately got rid of, and what you see be not a mere bag, without intestine or other organ: but only for the time being. For hear it, worn-out epicures, and old Indians who bemoan your livers, this little Holothuria knows a secret which, if he could tell it, you would be glad to buy of him for thousands sterling. To him blue pill and muriatic acid are superfluous, and travels to German Brunnen a waste of time. Happy Holothuria! who possesses really the secret of everlasting youth, which ancient fable bestowed on the serpent and the eagle. For when his teeth ache, or his digestive organs trouble him, all he has to do is just to cast up forthwith his entire inside, and, faisant maigre for a month or so, grow a fresh set, and then eat away as merrily as ever. His name, if you wish to consult so triumphant a hygeist, is Cucumaria Pentactes: but he has many a stout cousin round the Scotch coast, who knows the antibilious panacea as well as he, and submits, among the northern fishermen, to the rather rude and undeserved name of sea-puddings; one of which grows in Shetland to the enormous length of three feet, rivalling there his huge congeners, who display their exquisite plumes on every tropic coral reef. [9]

Next, what are those bright little buds, like salmon-coloured Banksia roses half expanded, sitting closely on the stone? Touch them; the soft part is retracted, and the orange flower of flesh is transformed into a pale pink flower of stone. That is the Madrepore, Caryophyllia Smithii (Plate V. fig. 2); one of our south coast rarities: and see, on the lip of the last one, which we have carefully scooped off with the chisel, two little pink towers of stone, delicately striated; drop them into this small bottle of sea-water, and from the top of each tower issues every half-second - what shall we call it? - a hand or a net of finest hairs, clutching at something invisible to our grosser sense. That is the Pyrgoma, parasitic only (as far as we know) on the lip of this same rare Madrepore; a little "cirrhipod," the cousin of those tiny barnacles which roughen every rock (a larger sort whereof I showed you on the Turritella), and of those larger ones also who burrow in the thick hide of the whale, and, borne about upon his mighty sides, throw out their tiny casting nets, as this Pyrgoma does, to catch every passing animalcule, and sweep them into the jaws concealed within its shell. And this creature, rooted to one spot through life and death, was in its infancy a free swimming animal, hovering from place to place upon delicate ciliae, till, having sown its wild oats, it settled down in life, built itself a good stone house, and became a landowner, or rather a glebae adscriptus, for ever and a day. Mysterious destiny! - yet not so mysterious as that of the free medusoid young of every polype and coral, which ends as a rooted tree of horn or stone, and seems to the eye of sensuous fancy to have literally degenerated into a vegetable. Of them you must read for yourself in Mr. Gosse's book; in the meanwhile he shall tell you something of the beautiful Madrepores themselves. His description, [10] by far the best yet published, should be read in full; we must content ourselves with extracts.

"Doubtless you are familiar with the stony skeleton of our Madrepore, as it appears in museums. It consists of a number of thin calcareous plates standing up edgewise, and arranged in a radiating manner round a low centre. A little below the margin their individuality is lost in the deposition of rough calcareous matter . . . The general form is more or less cylindrical, commonly wider at top than just above the bottom . . . This is but the skeleton; and though it is a very pretty object, those who are acquainted with it alone, can form but a very poor idea of the beauty of the living animal . . . Let it, after being torn from the rock, recover its equanimity; then you will see a pellucid gelatinous flesh emerging from between the plates, and little exquisitely formed and coloured tentacula, with white clubbed tips fringing the sides of the cup-shaped cavity in the centre, across which stretches the oval disc marked with a star of some rich and brilliant colour, surrounding the central mouth, a slit with white crenated lips, like the orifice of one of those elegant cowry shells which we put upon our mantelpieces. The mouth is always more or less prominent, and can be protruded and expanded to an astonishing extent. The space surrounding the lips is commonly fawn colour, or rich chestnut-brown; the star or vandyked circle rich red, pale vermilion, and sometimes the most brilliant emerald green, as brilliant as the gorget of a humming-bird."

And what does this exquisitely delicate creature do with its pretty mouth? Alas for fact! It sips no honey-dew, or fruits from paradise. - "I put a minute spider, as large as a pin's head, into the water, pushing it down to the coral. The instant it touched the tip of a tentacle, it adhered, and was drawn in with the surrounding tentacles between the plates. With a lens I saw the small mouth slowly open, and move over to that side, the lips gaping unsymmetrically; while with a movement as imperceptible as that of the hour hand of a watch, the tiny prey was carried along between the plates to the corner of the mouth. The mouth, however, moved most, and at length reached the edges of the plates, gradually closed upon the insect, and then returned to its usual place in the centre."

Mr. Gosse next tried the fairy of the walking mouth with a house- fly, who escaped only by hard fighting; and at last the gentle creature, after swallowing and disgorging various large pieces of shell-fish, found viands to its taste in "the lean of cooked meat and portions of earthworms," filling up the intervals by a perpetual dessert of microscopic animalcules, whirled into that lovely avernus, its mouth, by the currents of the delicate ciliae which clothe every tentacle. The fact is, that the Madrepore, like those glorious sea-anemones whose living flowers stud every pool, is by profession a scavenger and a feeder on carrion; and being as useful as he is beautiful, really comes under the rule which he seems at first to break, that handsome is who handsome does.

Another species of Madrepore [11] was discovered on our Devon coast by Mr. Gosse, more gaudy, though not so delicate in hue as our Caryophyllia. Mr. Gosse's locality, for this and numberless other curiosities, is Ilfracombe, on the north coast of Devon. My specimens came from Lundy Island, in the mouth of the Bristol Channel, or more properly from that curious "Rat Island" to the south of it, where still lingers the black long-tailed English rat, exterminated everywhere else by his sturdier brown cousin of the Hanoverian dynasty.

Look, now, at these tiny saucers of the thinnest ivory, the largest not bigger than a silver threepence, which contain in their centres a milk-white crust of stone, pierced, as you see under the magnifier, into a thousand cells, each with its living architect within. Here are two kinds: in one the tubular cells radiate from the centre, giving it the appearance of a tiny compound flower, daisy or groundsel; in the other they are crossed with waving grooves, giving the whole a peculiar fretted look, even more beautiful than that of the former species. They are Tubulipora patina and Tubulipora hispida; - and stay - break off that tiny rough red wart, and look at its cells also under the magnifier: it is Cellepora pumicosa; and now, with the Madrepore, you hold in your hand the principal, at least the commonest, British types of those famed coral insects, which in the tropics are the architects of continents, and the conquerors of the ocean surge. All the world, since the publication of Darwin's delightful "Voyage of the Beagle,"' and of Williams' "Missionary Enterprises," knows, or ought to know, enough about them: for those who do not, there are a few pages in the beginning of Dr. Landsborough's "British Zoophytes," well worth perusal.

There are a few other true cellepore corals round the coast. The largest of all, Cervicornis, may be dredged a few miles outside on the Exmouth bank, with a few more Tubulipores: but all tiny things, the lingering and, as it were, expiring remnants of that great coral-world which, through the abysmal depths of past ages, formed here in Britain our limestone hills, storing up for generations yet unborn the materials of agriculture and architecture. Inexpressibly interesting, even solemn, to those who will think, is the sight of those puny parasites which, as it were, connect the ages and the aeons: yet not so solemn and full of meaning as that tiny relic of an older world, the little pear-shaped Turbinolia (cousin of the Madrepores and Sea-anemones), found fossil in the Suffolk Crag, and yet still lingering here and there alive in the deep water of Scilly and the west coast of Ireland, possessor of a pedigree which dates, perhaps, from ages before the day in which it was said, "Let us make man in our image, after our likeness." To think that the whole human race, its joys and its sorrows, its virtues and its sins, its aspirations and its failures, has been rushing out of eternity and into eternity again, as Arjoon in the Bhagavad Gita beheld the race of men issuing from Kreeshna's flaming mouth, and swallowed up in it again, "as the crowds of insects swarm into the flame, as the homeless streams leap down into the ocean bed," in an everlasting heart-pulse whose blood is living souls - and all that while, and ages before that mystery began, that humble coral, unnoticed on the dark sea-floor, has been "continuing as it was at the beginning," and fulfilling "the law which cannot be broken," while races and dynasties and generations have been

> "Playing such fantastic tricks before high heaven,
> As make the angels weep."

Yes; it is this vision of the awful permanence and perfection of the natural world, beside the wild flux and confusion, the mad struggles, the despairing cries of the world of spirits which man has defiled by sin, which would at moments crush the naturalist's heart, and make his brain swim with terror, were it not that he can see by faith, through all the abysses and the ages, not merely

" Hands,
From out the darkness, shaping man;"

but above them a living loving countenance, human and yet Divine; and can hear a voice which said at first, "Let us make man in our image;" and hath said since then, and says for ever and for ever, "Lo, I am with you alway, even to the end of the world."

But now, friend, who listenest, perhaps instructed, and at least amused - if, as Professor Harvey well says, the simpler animals represent, as in a glass, the scattered organs of the higher races, which of your organs is represented by that "sca'd man's head," which the Devon children more gracefully, yet with less adherence to plain likeness, call "mermaid's head," [12] which we picked up just now on Paignton Sands? Or which, again, by its more beautiful little congener, [13] five or six of which are adhering tightly to the slab before us, a ball covered with delicate spines of lilac and green, and stuck over (cunning fellows!) with stripes of dead sea-weed to serve as improvised parasols? One cannot say that in him we have the first type of the human skull: for the resemblance, quaint as it is, is only sensuous and accidental, (in the logical use of that term,) and not homological, I.E. a lower manifestation of the same idea. Yet how is one tempted to say, that this was Nature's first and lowest attempt at that use of hollow globes of mineral for protecting soft fleshy parts, which she afterwards developed to such perfection in the skulls of vertebrate animals! But even that conceit, pretty as it sounds, will not hold good; for though Radiates similar to these were among the earliest tenants of the abyss, yet as early as their time, perhaps even before them, had been conceived and actualized, in the sharks, and in Mr. Hugh Miller's pets the old red sandstone fishes, that very true vertebrate skull and brain, of which this is a mere mockery. [14] Here the whole animal, with his extraordinary feeding mill, (for neither teeth nor jaws is a fit word for it,) is enclosed within an ever-growing limestone castle, to the architecture of which the Eddystone and the Crystal Palace are bungling heaps; without arms or legs, eyes or ears, and yet capable, in spite of his perpetual imprisonment, of walking, feeding, and breeding, doubt it not, merrily enough. But this result has been attained at the expense of a complication of structure, which has baffled all human analysis and research into final causes. As much concerning this most miraculous of families as is needful to be known, and ten times more than you are likely to understand, may be read in Harvey's "Sea-Side Book," pp. 142–148, - pages from which you will probably arise with a sense of the infinity and complexity of Nature, even in what we are pleased to call her "lower" forms, and the simplest and, as it were, easiest forms of life. Conceive a Crystal Palace, (for mere difference in size, as both the naturalist and the metaphysician know, has nothing to do with the wonder,) whereof each separate joist, girder, and pane grows continually without altering the shape of the whole; and you have conceived only one of the miracles embodied in that little sea-egg, which the Creator has, as it were, to justify to man His own immutability, furnished with a shell capable of enduring fossil for countless ages, that we may confess Him to have been as great when first His Spirit brooded on the deep, as He is now and will be through all worlds to come.

But we must make haste; for the tide is rising fast, and our stone will be restored to its eleven hours' bath, long before we have talked over half the wonders which it holds. Look though, ere you retreat, at one or two more.

What is that little brown thing whom you have just taken off the rock to which it adhered so stoutly by his sucking-foot? A limpet? Not at all: he is of quite a different family and structure; but, on the whole, a limpet-like shell would suit him well enough, so he had one given him: nevertheless, owing to certain anatomical peculiarities, he needed one aperture more than a limpet; so one, if you will examine, has been given him at the top of his shell. [15] This is one instance among a thousand of the way in which a scientific knowledge of objects must not obey, but run counter to, the impressions of sense; and of a custom in nature which makes this caution so necessary, namely, the repetition of the same form, slightly modified, in totally different animals, sometimes as if to avoid waste, (for why should not the same conception be used in two different cases, if it will suit in both?) and sometimes (more marvellous by far) when an organ, fully developed and useful in one species, appears in a cognate species but feeble, useless, and, as it were, abortive; and gradually, in species still farther removed, dies out altogether; placed there, it would seem, at first sight, merely to keep up the family likeness. I am half jesting; that cannot be the only reason, perhaps not the reason at all; but the fact is one of the most curious, and notorious also, in comparative anatomy.

Look, again, at those sea-slugs. One, some three inches long, of a bright lemon-yellow, clouded with purple; another of a dingy grey; [16] another exquisite little creature of a pearly French White, [17] furred all over the back with what seem arms, but are really gills, of ringed white and grey and black. Put that yellow one into water, and from his head, above the eyes, arise two serrated horns, while from the after-part of his back springs a circular Prince-of-Wales's-feather of gills, - they are almost exactly like those which we saw just now in the white Cucumaria. Yes; here is another instance of the same custom of repetition. The Cucumaria is a low radiate animal - the sea-slug a far higher mollusc; and every organ within him is formed on a different type; as indeed are those seemingly identical gills, if you come to examine them under the microscope, having to oxygenate fluids of a very different and more complicated kind; and, moreover, the Cucumaria's gills were put round his mouth, the Doris's feathers round the other extremity; that grey Eolis's, again, are simple clubs, scattered over his whole back, and in each of his nudibranch congeners these same gills take some new and fantastic form; in Melibaea those clubs are covered with warts; in Scyllaea, with tufted bouquets; in the beautiful Antiopa they are transparent bags; and in many other English species they take every conceivable form of leaf, tree, flower, and branch, bedecked with every colour of the rainbow, as you may see them depicted in Messrs. Alder and Hancock's unrivalled Monograph on the Nudibranch Mollusca.

And now, worshipper of final causes and the mere useful in nature, answer but one question, - Why this prodigal variety? All these Nudibranchs live in much the same way: why would not the same mould have done for them all? And why, again, (for we must push the argument a little further,) why have not all the butterflies, at least all who feed on the same plant, the same markings? Of all unfathomable triumphs of design, (we can only express ourselves thus, for honest induction, as Paley so well teaches, allows us to ascribe such results only to the design of some personal will and mind,) what surpasses that by which the scales on a butterfly's wing are arranged to produce a certain pattern of artistic beauty beyond all painter's skill? What a waste of power, on any utilitarian theory of nature! And once more, why are those strange microscopic atomies, the Diatomaceae and Infusoria, which fill every stagnant pool; which fringe every branch of sea-weed; which form banks hundreds of miles long on the Arctic sea-floor, and the strata of whole moorlands; which pervade in millions the mass of every iceberg, and float aloft in countless swarms amid the clouds of the volcanic dust; - why are their tiny shells of flint as fantastically various in their quaint mathematical symmetry, as they are countless beyond the wildest dreams of the Poet? Mystery inexplicable on the conceited notion which, making man forsooth the centre of the universe, dares to believe that this variety of forms has existed for countless ages in abysmal sea-depths and untrodden forests, only that some few individuals of the Western races might, in these latter days, at last discover and admire a corner here and there of the boundless realms of beauty. Inexplicable, truly, if man be the centre and the object of their existence; explicable enough to him who believes that God has created all things for Himself, and rejoices in His own handiwork, and that the material universe is, as the wise man says, "A platform whereon His Eternal Spirit sports and makes melody." Of all the blessings which the study of nature brings to the patient observer, let none, perhaps, be classed higher than this: that the further he enters into those fairy gardens of life and birth, which Spenser saw and described in his great poem, the more he learns the awful and yet most comfortable truth, that they do not belong to him, but to One greater, wiser, lovelier than he; and as he stands, silent with awe, amid the pomp of Nature's ever-busy rest, hears, as of old, "The Word of the Lord God walking among the trees of the garden in the cool of the day."

One sight more, and we have done. I had something to say, had time permitted, on the ludicrous element which appears here and there in nature. There are animals, like monkeys and crabs, which seem made to be laughed at; by those at least who possess that most indefinable of faculties, the sense of the ridiculous. As long as man possesses muscles especially formed to enable him to laugh, we have no right to suppose (with some) that laughter is an accident of our fallen nature; or to find (with others) the primary cause of the ridiculous in the perception of unfitness or disharmony. And yet we shrink (whether rightly or wrongly, we can hardly tell) from attributing a sense of the ludicrous to the Creator of these forms. It may be a weakness on my part; at least I will hope it is a reverent one: but till we can find something corresponding to what we conceive of the Divine Mind in any class of phenomena, it is perhaps better not to talk about them at all, but observe a stoic "epoche," waiting for more light, and yet confessing that our own laughter is uncontrollable, and therefore we hope not unworthy of us, at many a strange creature and strange doing which we meet, from the highest ape to the lowest polype.

But, in the meanwhile, there are animals in which results so strange, fantastic, even seemingly horrible, are produced, that fallen man may be pardoned, if he shrinks from them in disgust. That, at least, must be a consequence of our own wrong state; for everything is beautiful and perfect in its place. It may be answered, "Yes, in its place; but its place is not yours. You had no business to look at it, and must pay the penalty for intermeddling." I doubt that answer; for surely, if man have liberty to do anything, he has liberty to search out freely his heavenly Father's works; and yet every one seems to have his antipathic animal; and I know one bred from his childhood to zoology by land and sea, and bold in asserting, and honest in feeling, that all without exception is beautiful, who yet cannot, after handling and petting and admiring all day long every uncouth and venomous beast, avoid a paroxysm of horror at the sight of the common house-spider. At all events, whether we were intruding or not, in turning this stone, we must pay a fine for having done so; for there lies an animal as foul and monstrous to the eye as "hydra, gorgon, or chimaera dire," and yet so wondrously fitted to its work, that we must needs endure for our own instruction to handle and to look at it. Its name, if you wish for it, is Nemertes; probably N. Borlasii; [18] a worm of very "low" organization, though well fitted enough for its own work. You see it? That black, shiny, knotted lump among the gravel, small enough to be taken up in a dessert spoon. Look now, as it is raised and its coils drawn out. Three feet - six - nine, at least: with a capability of seemingly endless expansion; a slimy tape of living caoutchouc, some eighth of an inch in diameter, a dark chocolate-black, with paler longitudinal lines. Is it alive? It hangs, helpless and motionless, a mere velvet string across the hand. Ask the neighbouring Annelids and the fry of the rock fishes, or put it into a vase at home, and see. It lies motionless, trailing itself among the gravel; you cannot tell where it begins or ends; it may be a dead strip of sea-weed, Himanthalia lorea, perhaps, or Chorda filum; or even a tarred string. So thinks the little fish who plays over and over it, till he touches at last what is too surely a head. In an instant a bell-shaped sucker mouth has fastened to his side. In another instant, from one lip, a concave double proboscis, just like a tapir's (another instance of the repetition of forms), has clasped him like a finger; and now begins the struggle: but in vain. He is being "played" with such a fishing-line as the skill of a Wilson or a Stoddart never could invent; a living line, with elasticity beyond that of the most delicate fly-rod, which follows every lunge, shortening and lengthening, slipping and twining round every piece of gravel and stem of sea-weed, with a tiring drag such as no Highland wrist or step could ever bring to bear on salmon or on trout. The victim is tired now; and slowly, and yet dexterously, his blind assailant is feeling and shifting along his side, till he reaches one end of him; and then the black lips expand, and slowly and surely the curved finger begins packing him end-foremost down into the gullet, where he sinks, inch by inch, till the swelling which marks his place is lost among the coils, and he is probably macerated to a pulp long before he has reached the opposite extremity of his cave of doom. Once safe down, the black murderer slowly contracts again into a knotted heap, and lies, like a boa with a stag inside him, motionless and blest. [19]

There; we must come away now, for the tide is over our ankles; but touch, before you go, one of those little red mouths which peep out of the stone. A tiny jet of water shoots up almost into your face.

The bivalve [20] who has burrowed into the limestone knot (the softest part of the stone to his jaws, though the hardest to your chisel) is scandalized at having the soft mouths of his siphons so rudely touched, and taking your finger for some bothering Annelid, who wants to nibble him, is defending himself; shooting you, as naturalists do humming-birds, with water. Let him rest in peace; it will cost you ten minutes' hard work, and much dirt, to extract him; but if you are fond of shells, secure one or two of those beautiful pink and straw-coloured scallops (Hinnites pusio, Plate X. fig. 1), who have gradually incorporated the layers of their lower valve with the roughnesses of the stone, destroying thereby the beautiful form which belongs to their race, but not their delicate colour. There are a few more bivalves too, adhering to the stone, and those rare ones, and two or three delicate Mangeliae and Nassae [21] are trailing their graceful spires up and down in search of food. That little bright red and yellow pea, too, touch it - the brilliant coloured cloak is withdrawn, and, instead, you have a beautiful ribbed pink cowry, [22] our only European representative of that grand tropical family. Cast one wondering glance, too, at the forest of zoophytes and corals, Lepraliae and Flustrae, and those quaint blue stars, set in brown jelly, which are no zoophytes, but respectable molluscs, each with his well- formed mouth and intestines, [23] but combined in a peculiar form of Communism, of which all one can say is, that one hopes they like it; and that, at all events, they agree better than the heroes and heroines of Mr. Hawthorne's "Blithedale Romance."

Now away, and as a specimen of the fertility of the water-world, look at this rough list of species, [24] the greater part of which are on this very stone, and all of which you might obtain in an hour, would the rude tide wait for zoologists: and remember that the number of individuals of each species of polype must be counted by tens of thousands; and also, that, by searching the forest of sea-weeds which covers the upper surface, we should probably obtain some twenty minute species more.

A goodly catalogue this, surely, of the inhabitants of three or four large stones; and yet how small a specimen of the multitudinous nations of the sea!

From the bare rocks above high-water mark, down to abysses deeper than ever plummet sounded, is life, everywhere life; fauna after fauna, and flora after flora, arranged in zones, according to the amount of light and warmth which each species requires, and to the amount of pressure which they are able to endure. The crevices of the highest rocks, only sprinkled with salt spray in spring-tides and high gales, have their peculiar little univalves, their crisp lichen-like sea-weed, in myriads; lower down, the region of the Fuci (bladder-weeds) has its own tribes of periwinkles and limpets;

below again, about the neap-tide mark, the region of the corallines and Algae furnishes food for yet other species who graze on its watery meadows; and beneath all, only uncovered at low spring-tide, the zone of the Laminariae (the great tangles and ore-weeds) is most full of all of every imaginable form of life. So that as we descend the rocks, we may compare ourselves (likening small things to great) to those who, descending the Andes, pass in a single day from the vegetation of the Arctic zone to that of the Tropics. And here and there, even at half-tide level, deep rock-basins, shaded from the sun and always full of water, keep up in a higher zone the vegetation of a lower one, and afford in nature an analogy to those deep "barrancos" which split the high table-land of Mexico, down whose awful cliffs, swept by cool sea-breezes, the traveller looks from among the plants and animals of the temperate zone, and sees far below, dim through their everlasting vapour-bath of rank hot steam, the mighty forms and gorgeous colours of a tropic forest.

"I do not wonder," says Mr. Gosse, in his charming "Naturalist's Rambles on the Devonshire Coast" (p. 187), "that when Southey had an opportunity of seeing some of those beautiful quiet basins hollowed in the living rock, and stocked with elegant plants and animals, having all the charm of novelty to his eye, they should have moved his poetic fancy, and found more than one place in the gorgeous imagery of his Oriental romances. Just listen to him

> "It was a garden still beyond all price,
> Even yet it was a place of paradise;
> And here were coral bowers,
> And grots of madrepores,
> And banks of sponge, as soft and fair to eye
> As e'er was mossy bed
> Whereon the wood-nymphs lie
> With languid limbs in summer's sultry hours.
> Here, too, were living flowers,
> Which, like a bud compacted,
> Their purple cups contracted;
> And now in open blossom spread,
> Stretch'd, like green anthers, many a seeking head.
> And arborets of jointed stone were there,

And plants of fibres fine as silkworm's thread;
Yea, beautiful as mermaid's golden hair
Upon the waves dispread.
Others that, like the broad banana growing,
Raised their long wrinkled leaves of purple hue,
Like streamers wide outflowing.' - KEHAMA, xvi. 5.

"A hundred times you might fancy you saw the type, the very original of this description, tracing, line by line, and image by image, the details of the picture; and acknowledging, as you proceed, the minute truthfulness with which it has been drawn. For such is the loveliness of nature in these secluded reservoirs, that the accomplished poet, when depicting the gorgeous scenes of Eastern mythology - scenes the wildest and most extravagant that imagination could paint - drew not upon the resources of his prolific fancy for imagery here, but was well content to jot down the simple lineaments of Nature as he saw her in plain, homely England.

"It is a beautiful and fascinating sight for those who have never seen it before, to see the little shrubberies of pink coralline - 'the arborets of jointed stone' - that fringe those pretty pools. It is a charming sight to see the crimson banana-like leaves of the Delesseria waving in their darkest corners; and the purple fibrous tufts of Polysiphonia and Ceramia, 'fine as silkworm's thread.' But there are many others which give variety and impart beauty to these tide-pools. The broad leaves of the Ulva, finer than the finest cambric, and of the brightest emerald-green, adorn the hollows at the highest level, while, at the lowest, wave tiny forests of the feathery Ptilota and Dasya, and large leaves, cut into fringes and furbelows, of rosy Rhodymeniae. All these are lovely to behold; but I think I admire as much as any of them, one of the commonest of our marine plants, Chondrus crispus. It occurs in the greatest profusion on this coast, in every pool between tide-marks; and everywhere - except in those of the highest level, where constant exposure to light dwarfs the plant, and turns it of a dull umber-brown tint - it is elegant in form and brilliant in colour. The expanding fan-shaped fronds, cut into segments, cut, and cut again, make fine bushy tufts in a deep pool, and every segment of every frond reflects a flush of the most lustrous azure, like that of a tempered sword-blade." - GOSSE'S DEVONSHIRE COAST, pp. 187-189.

And the sea-bottom, also, has its zones, at different depths, and its peculiar forms in peculiar spots, affected by the currents and the nature of the ground, the riches of which have to be seen, alas! rather by the imagination than the eye; for such spoonfuls of the treasure as the dredge brings up to us, come too often rolled and battered, torn from their sites and contracted by fear, mere hints to us of what the populous reality below is like. Often, standing on the shore at low tide, has one longed to walk on and in under the waves, as the water-ousel does in the pools of the mountain burn, and see it all but for a moment; and a solemn beauty and meaning has invested the old Greek fable of Glaucus the fisherman: how eating of the herb which gave his fish strength to leap back into their native element, he was seized on the spot with a strange longing to follow them under the waves, and became for ever a companion of the fair semi-human forms with which the Hellenic poets peopled their sunny bays and firths, feeding "silent flocks" far below on the green Zostera beds, or basking with them on the sunny ledges in the summer noon, or wandering in the still bays on sultry nights amid the choir of Amphitrite and her sea-nymphs:-

> "Joining the bliss of the gods, as they waken the coves with their laughter,"

in nightly revels, whereof one has sung, -

> "So they came up in their joy; and before them the roll of the surges
> Sank, as the breezes sank dead, into smooth green foam-flecked marble
> Awed; and the crags of the cliffs, and the pines of the mountains, were silent.
> So they came up in their joy, and around them the lamps of the sea-nymphs,
> Myriad fiery globes, swam heaving and panting, and rainbows,
> Crimson, and azure, and emerald, were broken in star-showers, l ighting,

Far in the wine-dark depths of the crystal, the gardens of
     Nereus,
Coral, and sea-fan, and tangle, the blooms and the palms
     of the ocean.
So they went on in their joy, more white than the foam
     which they scattered,
Laughing and singing and tossing and twining; while, eager,
     the Tritons
Blinded with kisses their eyes, unreproved, and above them
     in worship
Fluttered the terns, and the sea-gulls swept past them on
     silvery pinions,
Echoing softly their laughter; around them the wantoning
     dolphins
Sighed as they plunged, full of love; and the great sea-
     horses which bore them
Curved up their crests in their pride to the delicate arms of
     their riders,
Pawing the spray into gems, till a fiery rainfall, unharming,
Sparkled and gleamed on the limbs of the maids, and the
     coils of the mermen.
So they went on in their joy, bathed round with the fiery
     coolness,
Needing nor sun nor moon, self-lighted, immortal: but
     others,
Pitiful, floated in silence apart; on their knees lay the sea-
     boys
Whelmed by the roll of the surge, swept down by the anger
     of Nereus;
Hapless, whom never again upon quay or strand shall their
     mothers
Welcome with garlands and vows to the temples; but,
     wearily pining,
Gaze over island and main for the sails which return not;
     they, heedless,
Sleep in soft bosoms for ever, and dream of the surge and
     the sea- maids.
So they passed by in their joy, like a dream, on the murmuring
     ripple."

Such a rhapsody may be somewhat out of order, even in a popular scientific book; and yet one cannot help at moments envying the old Greek imagination, which could inform the soulless sea-world with a human life and beauty. For, after all, star-fishes and sea- anemones are dull substitutes for Sirens and Tritons; the lamps of the sea-nymphs, those glorious phosphorescent medusae whose beauty Mr. Gosse sets forth so well with pen and pencil, are not as attractive as the sea-nymphs themselves would be; and who would not, like Menelaus, take the grey old man of the sea himself asleep upon the rocks, rather than one of his seal-herd, probably too with the same result as the world-famous combat in the Antiquary, between Hector and Phoca? And yet - is there no human interest in these pursuits, more humanity and more divine, than there would be even in those Triton and Nereid dreams, if realized to sight and sense? Heaven forbid that those should say so, whose wanderings among rock and pool have been mixed up with holiest passages of friendship and of love, and the intercommunion of equal minds and sympathetic hearts, and the laugh of children drinking in health from every breeze and instruction at every step, running ever and anon with proud delight to add their little treasure to their parents' stock, and of happy friendly evenings spent over the microscope and the vase, in examining, arranging, preserving, noting down in the diary the wonders and the labours of the happy, busy day. No; such short glimpses of the water-world as our present appliances afford us are full enough of pleasure; and we will not envy Glaucus: we will not even be over-anxious for the success of his only modern imitator, the French naturalist who is reported to have fitted himself with a waterproof dress and breathing apparatus, in order to walk the bottom of the Mediterranean, and see for himself how the world goes on at the fifty-fathom line: we will be content with the wonders of the shore and of the sea-floor, as far as the dredge will discover them to us. We shall even thus find enough to occupy (if we choose) our lifetime. For we must recollect that this hasty sketch has hardly touched on that vegetable water-world, which is as wonderful and as various as the animal one. A hint or two of the beauty of the sea- weeds has been given; but space has allowed no more. Yet we might have spent our time with almost as much interest and profit, had we neglected utterly the animals which we have found, and devoted our attention exclusively to the flora of the rocks. Sea-weeds are no mere playthings for children; and to buy at a shop some thirty pretty kinds, pasted on paper, with long names (probably mis-spelt) written under each, is not by any means to possess a collection of them. Putting aside the number and the obscurity of their species, the questions which arise in studying their growth, reproduction, and organic chemistry are of the very deepest and most important in the whole range of science; and it will need but a little study of such a book as Harvey's "Algae," to show the wise man that he who has comprehended (which no man yet does) the mystery of a single spore or tissue-cell, has reached depths in the great "Science of Life" at which an Owen would still confess himself "blind by excess of light." "Knowest thou how the bones grow in the womb?" asks the Jewish sage, sadly, half self-reprovingly, as he discovers that man is not the measure of all things, and that in much learning may be vanity and vexation of spirit, and in much study a weariness of the flesh; and all our deeper physical science only brings the same question more awfully near. "Vilior algâ," more worthless than the very sea-weed, says the old Roman: and yet no torn scrap of that very sea-weed, which to-morrow may manure the nearest garden, but says to us, "Proud man! talking of spores and vesicles, if thou darest for a moment to fancy that to have seen spores and vesicles is to have seen me, or to know what I am, answer this. Knowest thou how the bones do grow in the womb? Knowest thou even how one of these tiny black dots, which thou callest spores, grow on my fronds?" And to that question what answer shall we make? We see tissues divide, cells develop, processes go on - but How and Why? These are but phenomena; but what are phenomena save effects? Causes, it may be, of other effects; but still effects of other causes. And why does the cause cause that effect? Why should it not cause something else? Why should it cause anything at all? Because it obeys a law. But why does it obey the law? and how does it obey the law? And, after all, what is a law? A mere custom of Nature. We see the same phenomenon happen a great many times; and we infer from thence that it has a custom of happening; and therefore we call it a law: but we have not seen the law; all we have seen is the phenomenon which we suppose to indicate the law. We have seen things fall: but we never saw a little flying thing pulling them down, with "gravitation" labelled on its back; and the question, why things fall, and HOW, is just where it was before Newton was born, and is likely to remain there. All we can say is, that Nature has her customs, and that other customs ensue, when those customs appear: but that as to what connects cause and effect, as to what is the reason, the final cause, or even the CAUSA CAUSANS, of any phenomenon, we know not more but less than ever; for those laws or customs which seem to us simplest ("endosmose," for instance, or "gravitation"), are just the most inexplicable, logically unexpected, seemingly arbitrary, certainly supernatural - miraculous, if you will; for no natural and physical cause whatsoever can be assigned for them; while if anyone shall argue against their being miraculous and supernatural on the ground of their being so common, I can only answer, that of all absurd and illogical arguments, this is the most so. For what has the number of times which the miracle occurs to do with the question, save to increase the wonder? Which is more strange, that an inexplicable and unfathomable thing should occur once and for all, or that it should occur a million times every day all the world over?

Let those, however, who are too proud to wonder, do as seems good to them. Their want of wonder will not help them toward the required explanation: and to them, as to us, as soon as we begin asking, "HOW?" and "WHY?" the mighty Mother will only reply with that magnificent smile of hers, most genial, but most silent, which she has worn since the foundation of all worlds; that silent smile which has tempted many a man to suspect her of irony, even of deceit and hatred of the human race; the silent smile which Solomon felt, and answered in "Ecclesiastes;" which Goethe felt, and did not answer in his "Faust;" which Pascal felt, and tried to answer in his "Thoughts," and fled from into self-torture and superstition, terrified beyond his powers of endurance, as he found out the true meaning of St. John's vision, and felt himself really standing on that fragile and slippery "sea of glass," and close beneath him the bottomless abyss of doubt, and the nether fires of moral retribution. He fled from Nature's silent smile, as that poor old King Edward (mis-called the Confessor) fled from her hymns of praise, in the old legend of Havering-atte-bower, when he cursed the nightingales because their songs confused him in his prayers: but the wise man need copy neither, and fear neither the silence nor the laughter of the mighty mother Earth, if he will be but wise, and hear her tell him, alike in both - "Why call me mother? Why ask me for knowledge which I cannot teach, peace which I cannot give or take away? I am only your foster-mother and your nurse - and I have not been an unkindly one. But you are God's children, and not mine. Ask Him. I can amuse you with my songs; but they are but a nurse's lullaby to the weary flesh. I can awe you with my silence; but my silence is only my just humility, and your gain. How dare I pretend to tell you secrets which He who made me knows alone? I am but inanimate matter; why ask of me things which belong to living spirit? In God I live and move, and have my being; I know not how, any more than you know. Who will tell you what life is, save He who is the Lord of life? And if He will not tell you, be sure it is because you need not to know. At least, why seek God in nature, the living among the dead? He is not here: He is risen."

He is not here: He is risen. Good reader, you will probably agree that to know that saying, is to know the key-note of the world to come. Believe me, to know it, and all it means, is to know the keynote of this world also, from the fall of dynasties and the fate of nations, to the sea-weed which rots upon the beach. It may seem startling, possibly (though I hope not, for my readers' sake, irreverent), to go back at once after such thoughts, be they true or false, to the weeds upon the cliff above our heads. But He who is not here, but is risen, yet is here, and has appointed them their services in a wonderful order; and I wish that on some day, or on many days, when a quiet sea and offshore breezes have prevented any new objects from coming to land with the rising tide, you would investigate the flowers peculiar to our sea-rocks and sandhills. Even if you do not find the delicate lily-like Trichonema of the Channel Islands and Dawlish, or the almost as beautiful Squill of the Cornish cliffs, or the sea-lavender of North Devon, or any of those rare Mediterranean species which Mr. Johns has so charmingly described in his "Week at the Lizard Point," yet an average cliff, with its carpeting of pink thrift and of bladder catchfly, and Lady's finger, and elegant grasses, most of them peculiar to the sea marge, is often a very lovely flower- bed.

Not merely interesting, too, but brilliant in their vegetation are sandhills; and the seemingly desolate dykes and banks of salt marshes will yield many a curious plant, which you may neglect if you will: but lay to your account the having to repent your neglect hereafter, when, finding out too late what a pleasant study botany is, you search in vain for curious forms over which you trod every day in crossing flats which seemed to you utterly ugly and uninteresting, but which the good God was watching as carefully as He did the pleasant hills inland: perhaps even more carefully; for the uplands He has completed, and handed over to man, that he may dress and keep them: but the tide-flats below are still unfinished, dry land in the process of creation, to which every tide is adding the elements of fertility, which shall grow food, perhaps in some future state of our planet, for generations yet unborn.

But to return to the water-world, and to dredging; which of all sea-side pursuits is perhaps the most pleasant, combining as it does fine weather sailing with the discovery of new objects, to which, after all, the waifs and strays of the beach, whether "flotsom jetsom, or lagand," as the old Admiralty laws define them, are few and poor. I say particularly fine weather sailing; for a swell, which makes the dredge leap along the bottom, instead of scraping steadily, is as fatal to sport as it is to some people's comfort. But dredging, if you use a pleasure boat and the small naturalist's dredge, is an amusement in which ladies, if they will, may share, and which will increase, and not interfere with, the amusements of a water-party.

The naturalist's dredge, of which Mr. Gosse's "Aquarium" gives a detailed account, should differ from the common oyster dredge in being smaller; certainly not more than four feet across the mouth; and instead of having but one iron scraping-lip like the oyster dredge, it should have two, one above and one below, so that it will work equally well on whichsoever side it falls, or how often soever it may be turned over by rough ground. The bag-net should be of strong spunyarn, or (still better) of hide "such as those hides of the wild cattle of the Pampas, which the tobacconists receive from South America," cut into thongs, and netted close. It should be loosely laced together with a thong at the tail edge in order to be opened easily, when brought on board, without canting the net over, and pouring the contents roughly out through the mouth. The dragging-rope should be strong, and at least three times as long as the perpendicular depth of the water in which you are working; if, indeed, there is much breeze, or any swell at all, still more line should be veered out. The inboard end should be made fast somewhere in the stern sheets, the dredge hove to windward, the boat put before the wind; and you may then amuse yourself as you will for the next quarter of an hour, provided that you have got ready various wide-mouthed bottles for the more delicate monsters, and a couple of buckets, to receive the large lumps of oysters and serpulae which you will probably bring to the surface.

As for a dredging ground, one may be found, I suppose, off every watering-place. The most fertile spots are in rough ground, in not less than five fathoms water. The deeper the water, the rarer and more interesting will the animals generally be: but a greater depth than fifteen fathoms is not easily reached on this side of Plymouth; and, on the whole, the beginner will find enough in seven or eight fathoms to stock an aquarium rivalling any of those in the "Tank-house" at the Zoological Gardens.

In general, the south coast of England, to the eastward of Portland, affords bad dredging ground. The friable cliffs, of comparatively recent formations, keep the sea shallow, and the bottom smooth and bare, by the vast deposits of sand and gravel. Yet round the Isle of Wight, especially at the back of the Needles, there ought to be fertile spots; and Weymouth, according to Mr. Gosse and other well-known naturalists, is a very garden of Nereus. Torbay, as may well be supposed, is an admirable dredging spot; perhaps its two best points are round the isolated Thatcher and Oare-rock, and from the mouth of Brixham harbour to Berry Head; along which last line, for perhaps three hundred years, the decks of all Brixham trawlers have been washed down ere running into harbour, and the sea-bottom thus stored with treasures scraped up from deeper water in every direction for miles and miles.

Hastings is, I fear, but a poor spot for dredging. Its friable cliffs and strong tides produce a changeable and barren sea-floor. Yet the immense quantities of Flustra thrown up after a storm indicate dredging ground at no great distance outside; its rocks, uninteresting as they are compared with our Devonians, have yielded to the industry and science of M. Tumanowicz a vast number of sea- weeds and sponges. Those three curious polypes, Valkeria cuscuta (Plate I. fig. 3), Notamia Bursaria, and Serialaria Lendigera, abound within tide-marks; and as the place is so much visited by Londoners, it may be worth while to give a few hints as to what might be done, by anyone whose curiosity has been excited by the salt-water tanks of the Zoological Gardens and the Crystal Palace.

An hour or two's dredging round the rocks to the eastward, would probably yield many delicate and brilliant little fishes; Gobies, brilliant Labri, blue, yellow, and orange, with tiny rabbit mouths, and powerful protruding teeth; pipe fishes (Syngnathi) [25] with strange snipe-bills (which they cannot open) and snake-like bodies; small cuttlefish (Sepiolae) of a white jelly mottled with brilliant metallic hues, with a ring of suckered arms round their tiny parrots' beaks, who, put into a jar, will hover and dart in the water, as the skylark does in air, by rapid winnowings of their glassy side-fins, while they watch you with bright lizard-eyes; the whole animal being a combination of the vertebrate and the mollusc, so utterly fantastic and abnormal, that (had not the family been amongst the commonest, from the earliest geological epochs) it would have seemed, to man's deductive intellect, a form almost as impossible as the mermaid, far more impossible than the sea- serpent. These, and perhaps a few handsome sea-slugs and bivalve shells, you will be pretty sure to find: perhaps a great deal more.

Meanwhile, without dredging, you may find a good deal on the shore. In the spring Doris bilineata comes to the rocks in thousands, to lay its strange white furbelows of spawn upon their overhanging edges. Eolides of extraordinary beauty haunt the same spots. The great Eolis papillosa, of a delicate French grey; Eolis pellucida (?) (Plate X. fig. 4), in which each papilla on the back is beautifully coloured with a streak of pink, and tipped with iron blue; and a most fantastical yellow little creature, so covered with plumes and tentacles that the body is invisible, which I believe to be the Idalia aspersa of Alder and Hancock.

At the bottom of the rock pools, behind St. Leonard's baths, may be found hundreds of the snipe's feather Anemone (Sagartia troglodytes), of every line; from the common brown and grey snipe's feather kind, to the white-horned Hesperus, the orange-horned Aurora, and a rich lilac and crimson variety, which does not seem to agree with either the Lilacinia or Rubicunda of Gosse. A more beautiful living bouquet could hardly be seen, than might be made of the varieties of this single species, from this one place.

On the outside sands between the end of the Marina and the Martello tower, you may find, at very low tides, great numbers of a sand-tube, about three inches long, standing up out of the sand. I do not mean the tubes of the Terebella, so common in all sands, which are somewhat flexible, and have their upper end fringed with a ragged ring of sandy arms: those I speak of are straight and stiff, and ending in a point upward. Draw them out of the sand - they will offer some resistance - and put them into a vase of water; you will see the worm inside expand two delicate golden combs, just like old-fashioned back-hair combs, of a metallic lustre, which will astonish you. With these combs the worm seems to burrow head downward into the sand; but whether he always remains in that attitude I cannot say. His name is Pectinaria Belgica. He is an Annelid, or true worm, connected with the Serpulea and Sabellae of which I have spoken already, and holds himself in his case like them, by hooks and bristles set on each ring of his body. In confinement he will probably come out of his case and die; when you may dissect him at your leisure, and learn a great deal more about him thereby than (I am sorry to say) I know.

But if you have courage to run out fifteen or twenty miles to the Diamond, you may find really rare and valuable animals. There is a risk, of course, of being blown over to the coast of France, by a change of wind; there is a risk also of not being able to land at night on the inhospitable Hastings beach, and of sleeping, as best you can, on board: but in the long days and settled fine weather of summer, the trip, in a stout boat, ought to be a safe and a pleasant one.

On the Diamond you will find many, or most of those gay creatures which attract your eye in the central row of tanks at the Zoological Gardens: great twisted masses of Serpulae, [26] those white tubes of stone, from the mouth of which protrude pairs of rose-coloured or orange fans, flashing in, quick as light, the moment that your finger approaches them or your shadow crosses the water.

You will dredge, too, the twelve-rayed sun-star (Solaster papposa), with his rich scarlet armour; and more strange, and quite as beautiful, the bird's foot star (Palmipes membranaceus), which you may see crawling by its thousand sucking-feet in the Crystal Palace tanks, a pentagonal webbed bird's foot, of scarlet and orange shagreen. With him, most probably, will be a specimen of the great purple heart-urchin (Spatangus purpureus), clothed in pale lilac horny spines, and other Echinoderms, for which you must consult

Forbes's "British Star-fishes:" but perhaps the species among them which will interest you most, will be the common brittle-star (Ophiocoma rosula), of which a hundred or so, I can promise, shall come up at a single haul of the dredge, entwining their long spine-clad arms in a seemingly inextricable confusion of "kaleidoscope" patterns (thanks to Mr. Gosse for the one right epithet), purple and azure, fawn, brown, green, grey, white and crimson; as if a whole bed of China-asters should have first come to life, and then gone mad, and fallen to fighting. But pick out, one by one, specimens from the tangled mass, and you will agree that no China- aster is so fair as this living stone-flower of the deep, with its daisy-like disc, and fine long prickly arms, which never cease their graceful serpentine motion, and its colours hardly alike in any two specimens. Handle them not, meanwhile, too roughly, lest, whether modesty or in anger, they begin a desperate course of gradual suicide, and, breaking off arm after arm piecemeal, fling them indignantly at their tormentor. Along with these you will certainly obtain a few of that fine bivalve, the great Scallop, which you have seen lying on every fishmonger's counter in Hastings. Of these you must pick out those which seem dirtiest and most overgrown with parasites, and place them carefully in a jar of salt water, where they may not be rubbed; for they are worth your examination, not merely for the sake of that ring of gem-like eyes which borders their "cloak," lying along the extreme out edge of the shell as the valves are half open, but for the sake of the parasites outside: corallines of exquisite delicacy, Plumulariae and Sertulariae, dead men's hands (Alcyonia), lumps of white or orange jelly, which will protrude a thousand star-like polypes, and the Tubularia indivisa, twisted tubes of fine straw, which ought already to have puzzled you; for you may pick them up in considerable masses on the Hastings beach after a south-west gale, and think long over them before you determine whether the oat-like stems and spongy roots belong to an animal, or a vegetable. Animals they are, nevertheless, though even now you will hardly guess the fact, when you see at the mouth of each tube a little scarlet flower, connected with the pink pulp which fills the tube. For a further description of this largest and handsomest of our Hydroid Polypes, I must refer you to Johnston, or, failing him, to Landsborough; and go on, to beg you not to despise those pink, or grey, or white lumps of jelly, which will expand in salt water into exquisite sea-anemones, of quite different forms from any which we have found along the rocks. One of them will certainly be the Dianthus, [27] which will open into a furbelowed flower, furred with innumerable delicate tentacula; and in the centre a mouth of the most delicate orange, the size of the whole animal being perhaps eight inches high and five across. Perhaps it will be of a satiny grey, perhaps pale rose, perhaps pure white; whatever its colour, it is the very maiden queen of all the beautiful tribe, and one of the loveliest gems with which it has pleased God to bedeck this lower world.

These and much more you will find on the scallops, or even more plentifully on any lump of ancient oysters; and if you do not dredge, it would be well worth your while to make interest with the fish-monger for a few oyster lumps, put into water the moment they are taken out of the trawl. Divide them carefully, clear out the oysters with a knife, and put the shells into your aquarium, and you will find that an oyster at home is a very different thing from an oyster on a stall.

You ought, besides, to dredge many handsome species of shells, which you would never pick up along the beach; and if you are conchologizing in earnest, you must not forget to bring home a tin box of shell sand, to be washed and picked over in a dish at your leisure, or forget either to wash through a fine sieve, over the boat's side, any sludge and ooze which the dredge brings up. Many - I may say, hundreds - rare and new shells are found in this way, and in no other.

But if you cannot afford the expense of your own dredge and boat, and the time and trouble necessary to follow the occupation scientifically, yet every trawler and oyster-boat will afford you a tolerable satisfaction. Go on board one of these; and while the trawl is down, spend a pleasant hour or two in talking with the simple, honest, sturdy fellows who work it, from whom (if you are as fortunate as I have been for many a year past) you may get many a moving story of danger and sorrow, as well as many a shrewd practical maxim, and often, too, a living recognition of God, and the providence of God, which will send you home, perhaps, a wiser and more genial man. And when the trawl is hauled, wait till the fish are counted out, and packed away, and then kneel down and inspect (in a pair of Mackintosh leggings, and your oldest coat) the crawling heap of shells and zoophytes which remains behind about the decks, and you will find, if a landsman, enough to occupy you for a week to come. Nay, even if it be too calm for trawling, condescend to go out in a dingy, and help to haul some honest fellow's deep-sea lines and lobster-pots, and you will find more and stranger things about them than even fish or lobsters: though they, to him who has eyes to see, are strange enough.

I speak from experience; for it was not so very long ago that, in the north of Devon, I found sermons, not indeed in stones, but in a creature reputed among the most worthless of sea-vermin. I had been lounging about all the morning on the little pier, waiting, with the rest of the village, for a trawling breeze which would not come. Two o'clock was past, and still the red mainsails of the skiffs hung motionless, and their images quivered head downwards in the glassy swell,

> "As idle as a painted ship
> Upon a painted ocean."

It was neap-tide, too, and therefore nothing could be done among the rocks. So, in despair, finding an old coast-guard friend starting for his lobster-pots, I determined to save the old man's arms, by rowing him up the shore; and then paddled homeward again, under the high green northern wall, five hundred feet of cliff furred to the water's edge with rich oak woods, against whose base the smooth Atlantic swell died whispering, as if curling itself up to sleep at last within that sheltered nook, tired with its weary wanderings. The sun sank lower and lower behind the deer-park point; the white stair of houses up the glen was wrapped every moment deeper and deeper in hazy smoke and shade, as the light faded; the evening fires were lighted one by one; the soft murmur of the waterfall, and the pleasant laugh of children, and the splash of homeward oars, came clearer and clearer to the ear at every stroke: and as we rowed on, arose the recollection of many a brave and wise friend, whose lot was cast in no such western paradise, but rather in the infernos of this sinful earth, toiling even then amid the festering alleys of Bermondsey and Bethnal Green, to palliate death and misery which they had vainly laboured to prevent, watching the strides of that very cholera which they had been striving for years to ward off, now re-admitted in spite of all their warnings, by the carelessness, and laziness, and greed of sinful man. And as I thought over the whole hapless question of sanitary reform, proved long since a moral duty to God and man, possible, easy, even pecuniarily profitable, and yet left undone, there seemed a sublime irony, most humbling to man, in some of Nature's processes, and in the silent and unobtrusive perfection with which she has been taught to anticipate, since the foundation of the world, some of the loftiest discoveries of modern science, of which we are too apt to boast as if we had created the method by discovering its possibility. Created it? Alas for the pride of human genius, and the autotheism which would make man the measure of all things, and the centre of the universe! All the invaluable laws and methods of sanitary reform at best are but clumsy imitations of the unseen wonders which every animalcule and leaf have been working since the world's foundation; with this slight difference between them and us, that they fulfil their appointed task, and we do not.

The sickly geranium which spreads its blanched leaves against the cellar panes, and peers up, as if imploringly, to the narrow slip of sunlight at the top of the narrow alley, had it a voice, could tell more truly than ever a doctor in the town, why little Bessy sickened of the scarlatina, and little Johnny of the hooping-cough, till the toddling wee things who used to pet and water it were carried off each and all of them one by one to the churchyard sleep, while the father and mother sat at home, trying to supply by gin that very vital energy which fresh air and pure water, and the balmy breath of woods and heaths, were made by God to give; and how the little geranium did its best, like a heaven-sent angel, to right the wrong which man's ignorance had begotten, and drank in, day by day, the poisoned atmosphere, and formed it into fair green leaves, and breathed into the children's faces from every pore, whenever they bent over it, the life-giving oxygen for which their dulled blood and festered lungs were craving in vain; fulfilling God's will itself, though man would not, too careless or too covetous to see, after thousands of years of boasted progress, why God had covered the earth with grass, herb, and tree, a living and life-giving garment of perpetual health and youth.

It is too sad to think long about, lest we become very Heraclituses. Let us take the other side of the matter with

Democritus, try to laugh man out of a little of his boastful ignorance and self-satisfied clumsiness, and tell him, that if the House of Commons would but summon one of the little Paramecia from any Thames' sewer-mouth, to give his evidence before their next Cholera Committee, sanitary blue-books, invaluable as they are, would be superseded for ever and a day; and sanitary reformers would no longer have to confess, that they know of no means of stopping the smells which in past hot summers drove the members out of the House, and the judges out of Westminster Hall.

Nay, in the boat at the minute of which I have been speaking, silent and neglected, sat a fellow-passenger, who was a greater adept at removing nuisances than the whole Board of Health put together; and who had done his work, too, with a cheapness unparalleled; for all his good deeds had not as yet cost the State one penny. True, he lived by his business; so do other inspectors of nuisances: but Nature, instead of paying Maia Squinado, Esquire, some five hundred pounds sterling per annum for his labour, had contrived, with a sublime simplicity of economy which Mr. Hume might have envied and admired afar off, to make him do his work gratis, by giving him the nuisances as his perquisites, and teaching him how to eat them. Certainly (without going the length of the Caribs, who upheld cannibalism because, they said, it made war cheap, and precluded entirely the need of a commissariat), this cardinal virtue of cheapness ought to make Squinado an interesting object in the eyes of the present generation; especially as he was at that moment a true sanitary martyr, having, like many of his human fellow-workers, got into a fearful scrape by meddling with those existing interests, and "vested rights which are but vested wrongs," which have proved fatal already to more than one Board of Health. For last night, as he was sitting quietly under a stone in four fathoms water, he became aware (whether by sight, smell, or that mysterious sixth sense, to us unknown, which seems to reside in his delicate feelers) of a palpable nuisance somewhere in the neighbourhood; and, like a trusty servant of the public, turned out of his bed instantly and went in search; till he discovered, hanging among what he judged to be the stems of ore-weed (Laminaria), three or four large pieces of stale thornback, of most evil savour, and highly prejudicial to the purity of the sea, and the health of the neighbouring herrings. Happy Squinado! He needed not to discover the limits of his authority, to consult any lengthy Nuisances' Removal Act, with its clauses, and counter-clauses, and explanations of interpretations, and interpretations of explanations. Nature, who can afford to be arbitrary, because she is perfect, and to give her servants irresponsible powers, because she has trained them to their work, had bestowed on him and on his forefathers, as general health inspectors, those very summary powers of entrance and removal in the watery realms for which common sense, public opinion, and private philanthropy are still entreating vainly in the terrestrial realms; so finding a hole, in he went, and began to remove the nuisance, without "waiting twenty-four hours," "laying an information," "serving a notice," or any other vain delay. The evil was there, - and there it should not stay; so having neither cart nor barrow, he just began putting it into his stomach, and in the meanwhile set his assistants to work likewise. For suppose not, gentle reader, that Squinado went alone; in his train were more than a hundred thousand as good as he, each in his office, and as cheaply paid; who needed no cumbrous baggage train of force-pumps, hose, chloride of lime packets, whitewash, pails or brushes, but were every man his own instrument; and, to save expense of transit, just grew on Squinado's back. Do you doubt the assertion? Then lift him up hither, and putting him gently into that shallow jar of salt water, look at him through the hand-magnifier, and see how Nature is maxima in minimis.

There he sits, twiddling his feelers (a substitute, it seems, with crustacea for biting their nails when they are puzzled), and by no means lovely to look on in vulgar eyes; - about the bigness of a man's fist; a round-bodied, spindle-shanked, crusty, prickly, dirty fellow, with a villanous squint, too, in those little bony eyes, which never look for a moment both the same way. Never mind: many a man of genius is ungainly enough; and Nature, if you will observe, as if to make up to him for his uncomeliness, has arrayed him as Solomon in all his glory never was arrayed, and so fulfilled one of the proposals of old Fourier - that scavengers, chimney- sweeps, and other workers in disgusting employments, should be rewarded for their self-sacrifice in behalf of the public weal by some peculiar badge of honour, or laurel crown. Not that his crown, like those of the old Greek games, is a mere useless badge; on the contrary, his robe of state is composed of his fellow- servants. His whole back is covered with a little grey forest of branching hairs, fine as a spider's web, each branchlet carrying its little pearly ringed club, each club its rose-coloured polype, like (to quote Mr. Gosse's comparison) the unexpanded birds of the acacia. [28]

On that leg grows, amid another copse of the grey polypes, a delicate straw-coloured Sertularia, branch on branch of tiny double combs, each tooth of the comb being a tube containing a living flower; on another leg another Sertularia, coarser, but still beautiful; and round it again has trained itself, parasitic on the parasite, plant upon plant of glass ivy, bearing crystal bells, [29] each of which, too, protrudes its living flower; on another leg is a fresh species, like a little heather-bush of whitest ivory, [30] and every needle leaf a polype cell - let us stop before the imagination grows dizzy with the contemplation of those myriads of beautiful atomies. And what is their use? Each living flower, each polype mouth is feeding fast, sweeping into itself, by the perpetual currents caused by the delicate fringes upon its rays (so minute these last, that their motion only betrays their presence), each tiniest atom of decaying matter in the surrounding water, to convert it, by some wondrous alchemy, into fresh cells and buds, and either build up a fresh branch in their thousand-tenanted tree, or form an egg-cell, from whence when ripe may issue, not a fixed zoophyte, but a free swimming animal.

And in the meanwhile, among this animal forest grows a vegetable one of delicatest sea-weeds, green and brown and crimson, whose office is, by their everlasting breath, to reoxygenate the impure water, and render it fit once more to be breathed by the higher animals who swim or creep around.

Mystery of mysteries! Let us jest no more, - Heaven forgive us if we have jested too much on so simple a matter as that poor spider-crab, taken out of the lobster-pots, and left to die at the bottom of the boat, because his more aristocratic cousins of the blue and purple armour will not enter the trap while he is within.

I am not aware whether the surmise, that these tiny zoophytes help to purify the water by exhaling oxygen gas, has yet been verified. The infusorial animalcules do so, reversing the functions of animal life, and instead of evolving carbonic acid gas, as other animals do, evolve pure oxygen. So, at least, says Liebig, who states that he found a small piece of matchwood, just extinguished, burst out again into a flame on being immersed in the bubbles given out by these living atomies.

I myself should be inclined to doubt that this is the case with zoophytes, having found water in which they were growing (unless, of course, sea-weeds were present) to be peculiarly ready to become foul; but it is difficult to say whether this is owing to their deoxygenating the water while alive, like other animals, or to the fact that it is very rare to get a specimen of zoophyte in which a large number of the polypes have not been killed in the transit home, or at least so far knocked about, that (in the Anthozoa, which are far the most abundant) the polype - or rather living mouth, for it is little more - is thrown off to decay, pending the growth of a fresh one in the same cell.

But all the sea-weeds, in common with other vegetables, perform this function continually, and thus maintain the water in which they grow in a state fit to support animal life.

This fact - first advanced by Priestley and Ingenhousz, and though doubted by the great Ellis, satisfactorily ascertained by Professor Daubeny, Mr. Ward, Dr. Johnston, and Mr. Warrington - gives an answer to the question, which I hope has ere now arisen in the minds of some of my readers, -

How is it possible to see these wonders at home? Beautiful and instructive as they may be, can they be meant for any but dwellers by the sea-side? Nay more, even to them, must not the glories of the water-world be always more momentary than those of the rainbow, a mere Fata Morgana which breaks up and vanishes before the eyes? If there were but some method of making a miniature sea-world for a few days; much more of keeping one with us when far inland. -

This desideratum has at last been filled up; and science has shown, as usual, that by simply obeying Nature, we may conquer her, even so far as to have our miniature sea, of artificial salt-water, filled with living plants and sea-weeds, maintaining each other in perfect health, and each following, as far as is possible in a confined space, its natural habits.

To Dr. Johnston is due, as far as is known, the honour of the first accomplishment of this as of a hundred other zoological triumphs. As early as 1842, he proved to himself the vegetable nature of the common pink Coralline, which fringes every rock-pool, by keeping it for eight weeks in unchanged salt-water, without any putrefaction ensuing. The ground, of course, on which the proof rested in this case was, that if the coralline were, as had often been thought, a zoophyte, the water would become corrupt, and poisonous to the life of the small animals in the same jar; and that its remaining fresh argued that the coralline had re-oxygenated it from time to time, and was therefore a vegetable.

In 1850, Mr. Robert Warrington communicated to the Chemical Society the results of a year's experiments, "On the Adjustment of the Relations between the Animal and Vegetable Kingdoms, by which the Vital Functions of both are permanently maintained." The law which his experiments verified was the same as that on which Mr. Ward, in 1842, founded his invaluable proposal for increasing the purity of the air in large towns, by planting trees and cultivating flowers in rooms, THAT THE ANIMAL AND VEGETABLE RESPIRATIONS MIGHT COUNTERBALANCE EACH OTHER; the animal's blood being purified by the oxygen given off by the plants, the plants fed by the carbonic acid breathed out by the animals.

On the same principle, Mr. Warrington first kept, for many months, in a vase of unchanged water, two small gold fish and a plant of Vallisneria spiralis; and two years afterwards began a similar experiment with sea-water, weeds, and anemones, which were, at last, as successful as the former ones. Mr. Gosse had, in the meanwhile, with tolerable success begun a similar method, unaware of what Mr. Warrington had done; and now the beautiful and curious exhibition of fresh and salt water tanks in the Zoological Gardens in London, bids fair to be copied in every similar institution, and we hope in many private houses, throughout the kingdom.

To this subject Mr. Gosse's book, "The Aquarium," is principally devoted, though it contains, besides, sketches of coast scenery, in his usual charming style, and descriptions of rare sea-animals, with wise and goodly reflections thereon. One great object of interest in the book is the last chapter, which treats fully of the making and stocking these salt-water "Aquaria;" and the various beautifully coloured plates, which are, as it were, sketches from the interior of tanks, are well fitted to excite the desire of all readers to possess such gorgeous living pictures, if as nothing else, still as drawing-room ornaments, flower-gardens which never wither, fairy lakes of perpetual calm which no storm blackens, -

[Greek text which cannot be reproduced]

Those who have never seen one of them can never imagine (and neither Mr. Gosse's pencil nor my clumsy words can ever describe to them) the gorgeous colouring and the grace and delicacy of form which these subaqueous landscapes exhibit.

As for colouring, - the only bit of colour which I can remember even faintly resembling them (for though Correggio's Magdalene may rival them in greens and blues, yet even he has no such crimsons and purples) is the Adoration of the Shepherds, by that "prince of colorists" - Palma Vecchio, which hangs on the left-hand side of Lord Ellesmere's great gallery. But as for the forms, - where shall we see their like? Where, amid miniature forests as fantastic as those of the tropics, animals whose shapes outvie the wildest dreams of the old German ghost painters which cover the walls of the galleries of Brussels or Antwerp? And yet the uncouthest has some quaint beauty of its own, while most - the star-fishes and anemones, for example - are nothing but beauty. The brilliant plates in Mr. Gosse's "Aquarium" give, after all, but a meagre picture of the reality, as it may be seen in the tank-house at the Zoological Gardens; and as it may be seen also, by anyone who will follow carefully the directions given at the end of his book, stock a glass vase with such common things as he may find in an hour's search at low tide, and so have an opportunity of seeing how truly Mr. Gosse says, in his valuable preface, that -

> "The habits" (and he might well have added, the marvellous beauty) "of animals will never be thoroughly known till they are observed in detail. Nor is it sufficient to mark them with attention now and then; they must be closely watched, their various actions carefully noted, their behaviour under different circumstances, and especially those movements which seem to us mere vagaries, undirected by any suggestive motive or cause, well examined. A rich fruit of result, often new and curious and unexpected, will, I am sure, reward anyone who studies living animals in

this way. The most interesting parts, by far, of published Natural History are those minute, but graphic particulars, which have been gathered up by an attentive watching of individual animals."

Mr. Gosse's own books, certainly, give proof enough of this. We need only direct the reader to his exquisitely humorous account of the ways and works of a captive soldier-crab, [31] to show them how much there is to be seen, and how full Nature is also of that ludicrous element of which we spoke above. And, indeed, it is in this form of Natural History: not in mere classification, and the finding out of means, and quarrellings as to the first discovery of that beetle or this buttercup, - too common, alas! among mere closet-collectors, - "endless genealogies," to apply St. Paul's words by no means irreverently or fancifully, "which do but gender strife;" - not in these pedantries is that moral training to be found, for which we have been lauding the study of Natural History: but in healthful walks and voyages out of doors, and in careful and patient watching of the living animals and plants at home, with an observation sharpened by practice, and a temper calmed by the continual practice of the naturalist's first virtues - patience and perseverance.

Practical directions for forming an "Aquarium" may be found in Mr. Gosse's book bearing that name, at pp. 101, 255, ET SEQ.; and those who wish to carry out the notion thoroughly, cannot do better than buy his book, and take their choice of the many different forms of vase, with rockwork, fountains, and other pretty devices which he describes.

But the many, even if they have Mr. Gosse's book, will be rather inclined to begin with a small attempt; especially as they are probably half sceptical of the possibility of keeping sea-animals inland without changing the water. A few simple directions, therefore, will not come amiss here. They shall be such as anyone can put into practice, who goes down to stay in a lodging-house at the most cockney of watering-places.

Buy at any glass-shop a cylindrical glass jar, some six inches in diameter and ten high, which will cost you from three to four shillings; wash it clean, and fill it with clean salt-water, dipped out of any pool among the rocks, only looking first to see that there is no dead fish or other evil matter in the said pool, and that no stream from the land runs into it. If you choose to take the trouble to dip up the water over a boat's side, so much the better.

So much for your vase; now to stock it.

Go down at low spring-tide to the nearest ledge of rocks, and with a hammer and chisel chip off a few pieces of stone covered with growing sea-weed. Avoid the common and coarser kinds (fuci) which cover the surface of the rocks; for they give out under water a slime which will foul your tank: but choose the more delicate species which fringe the edges of every pool at low-water mark; the pink coralline, the dark purple ragged dulse (Rhodymenia), the Carrageen moss (Chondrus), and above all, the commonest of all, the delicate green Ulva, which you will see growing everywhere in wrinkled fan-shaped sheets, as thin as the finest silver-paper. The smallest bits of stone are sufficient, provided the sea-weeds have hold of them; for they have no real roots, but adhere by a small disc, deriving no nourishment from the rock, but only from the water. Take care, meanwhile, that there be as little as possible on the stone, beside the weed itself. Especially scrape off any small sponges, and see that no worms have made their twining tubes of sand among the weed-stems; if they have, drag them out; for they will surely die, and as surely spoil all by sulphuretted hydrogen, blackness, and evil smells.

Put your weeds into your tank, and settle them at the bottom; which last, some say, should be covered with a layer of pebbles: but let the beginner leave it as bare as possible; for the pebbles only tempt cross-grained annelids to crawl under them, die, and spoil all by decaying: whereas if the bottom of the vase is bare, you can see a sickly or dead inhabitant at once, and take him out (which you must do) instantly. Let your weeds stand quietly in the vase a day or two before you put in any live animals; and even then, do not put any in if the water does not appear perfectly clear: but lift out the weeds, and renew the water ere you replace them.

This is Mr. Gosse's method. But Mr. Lloyd, in his "Handbook to the Crystal Palace Aquarium," advises that no weed should be put into the tank. "It is better," he says, "to depend only on those which gradually and naturally appear on the rocks of the aquarium by the action of light, and which answer every chemical purpose." I should advise anyone intending to set up an aquarium, however small, to study what Mr. Lloyd says on this matter in pp. 17-19, and also in page 30, of his pamphlet; and also to go to the Crystal Palace Aquarium, and there see for himself the many beautiful species of sea-weeds which have appeared spontaneously in the tanks from unsuspected spores floating in the sea-water. On the other hand, Mr. Lloyd lays much stress on the necessity of arating the water, by keeping it in perpetual motion; a process not easy to be carried out in small aquaria; at least to that perfection which has been attained at the Crystal Palace, where the water is kept in continual circulation by steam-power. For a jar-aquarium, it will be enough to drive fresh air through the water every day, by means of a syringe.

Now for the live stock. In the crannies of every rock you will find sea-anemones (Actiniae); and a dozen of these only will be enough to convert your little vase into the most brilliant of living flower-gardens. There they hang upon the under side of the ledges, apparently mere rounded lumps of jelly: one is of dark purple dotted with green; another of a rich chocolate; another of a delicate olive; another sienna-yellow; another all but white. Take them from their rock; you can do it easily by slipping under them your finger-nail, or the edge of a pewter spoon. Take care to tear the sucking base as little as possible (though a small rent they will darn for themselves in a few days, easily enough, and drop them into a basket of wet sea-weed; when you get home turn them into a dish full of water and leave them for the night, and go to look at them to-morrow. What a change! The dull lumps of jelly have taken root and flowered during the night, and your dish is filled from side to side with a bouquet of chrysanthemums; each has expanded into a hundred-petalled flower, crimson, pink, purple, or orange; touch one, and it shrinks together like a sensitive plant, displaying at the root of the petals a ring of brilliant turquoise beads. That is the commonest of all the Actiniae (Mesembryanthemum); you may have him when and where you will: but if you will search those rocks somewhat closer, you will find even more gorgeous species than him. See in that pool some dozen large ones, in full bloom, and quite six inches across, some of them. If their cousins whom we found just now were like Chrysanthemums, these are like quilled Dahlias. Their arms are stouter and shorter in proportion than those of the last species, but their colour is equally brilliant. One is a brilliant blood-red; another a delicate sea-blue striped with pink; but most have the disc and the innumerable arms striped and ringed with various shades of grey and brown. Shall we get them? By all means if we can. Touch one. Where is he now? Gone? Vanished into air, or into stone? Not quite. You see that knot of sand and broken shell lying on the rock, where your Dahlia was one moment ago. Touch it, and you will find it leathery and elastic. That is all which remains of the live Dahlia. Never mind; get your finger into the crack under him, work him gently but firmly out, and take him home, and he will be as happy and as gorgeous as ever to-morrow.

Let your Actiniae stand for a day or two in the dish, and then, picking out the liveliest and handsomest, detach them once more from their hold, drop them into your vase, right them with a bit of stick, so that the sucking base is downwards, and leave them to themselves thenceforth.

These two species (Mesembryanthemum and Crassicornis) are quite beautiful enough to give a beginner amusement: but there are two others which are not uncommon, and of such exceeding loveliness, that it is worth while to take a little trouble to get them. The one is Dianthus, which I have already mentioned; the other Bellis, the sea-daisy, of which there is an excellent description and plates in Mr. Gosse's "Rambles in Devon," pp. 24 to 32.

It is common at Ilfracombe, and at Torquay; and indeed everywhere where there are cracks and small holes in limestone or slate rock. In these holes it fixes its base, and expands its delicate brown-grey star-like flowers on the surface: but it must be chipped out with hammer and chisel, at the expense of much dirt and patience; for the moment it is touched it contracts deep into the rock, and all that is left of the daisy flower, some two or three inches across, is a blue knot of half the size of a marble. But it will expand again, after a day or two of captivity, and will repay all the trouble which it has cost. Troglodytes may be found, as I have said already, in hundreds at Hastings, in similar situations to that of Bellis; its only token, when the tide is down, being a round dimple in the muddy sand which firs the lower cracks of rocks.

But you will want more than these anemones, both for your own amusement, and for the health of your tank. Microscopic animals will breed, and will also die; and you need for them some such scavenger as our poor friend Squinado, to whom you were introduced a few pages back. Turn, then, a few stones which lie piled on each other at extreme low-water mark, and five minutes' search will give you the very animal you want, - a little crab, of a dingy russet above, and on the under side like smooth porcelain. His back is quite flat, and so are his large angular fringed claws, which, when he folds them up, lie in the same plane with his shell, and fit neatly into its edges. Compact little rogue that he is, made especially for sidling in and out of cracks and crannies, he carries with him such an apparatus of combs and brushes as Isidor or Floris never dreamed of; with which he sweeps out of the seawater at every moment shoals of minute animalcules, and sucks them into his tiny mouth. Mr. Gosse will tell you more of this marvel, in his "Aquarium," p. 48.

Next, your sea-weeds, if they thrive as they ought to do, will sow their minute spores in millions around them; and these, as they vegetate, will form a green film on the inside of the glass, spoiling your prospect: you may rub it off for yourself, if you will, with a rag fastened to a stick; but if you wish at once to save yourself trouble, and to see how all emergencies in nature are provided for, you will set three or four live shells to do it for you, and to keep your subaqueous lawn close mown.

That last word is no figure of speech. Look among the beds of sea- weed for a few of the bright yellow or green sea-snails (Nerita), or Conical Tops (Trochus), especially that beautiful pink one spotted with brown (Ziziphinus), which you are sure to find about shaded rock-ledges at dead low tide, and put them into your aquarium. For the present, they will only nibble the green ulvae; but when the film of young weed begins to form, you will see it mown off every morning as fast as it grows, in little semicircular sweeps, just as if a fairy's scythe had been at work during the night.

And a scythe has been at work; none other than the tongue of the little shell-fish; a description of its extraordinary mechanism (too long to quote here, but which is well worth reading) may be found in Gosse's "Aquarium." [32]

A prawn or two, and a few minute star-fish, will make your aquarium complete; though you may add to it endlessly, as one glance at the salt-water tanks of the Zoological Gardens, and the strange and beautiful forms which they contain, will prove to you sufficiently.

You have two more enemies to guard against, dust, and heat. If the surface of the water becomes clogged with dust, the communication between it and the life-giving oxygen of the air is cut off; and then your animals are liable to die, for the very same reason that fish die in a pond which is long frozen over, unless a hole be broken in the ice to admit the air. You must guard against this by occasional stirring of the surface, or, as I have already said, by syringing and by keeping on a cover. A piece of muslin tied over will do; but a better defence is a plate of glass, raised on wire some half-inch above the edge, so as to admit the air. I am not sure that a sheet of brown paper laid over the vase is not the best of all, because that, by its shade, also guards against the next evil, which is heat. Against that you must guard by putting a curtain of muslin or oiled paper between the vase and the sun, if it be very fierce, or simply (for simple expedients are best) by laying a handkerchief over it till the heat is past. But if you leave your vase in a sunny window long enough to let the water get tepid, all is over with your pets. Half an hour's boiling may frustrate the care of weeks. And yet, on the other hand, light you must have, and you can hardly have too much. Some animals certainly prefer shade, and hide in the darkest crannies; and for them, if your aquarium is large enough, you must provide shade, by arranging the bits of stone into piles and caverns. But without light, your sea-weeds will neither thrive nor keep the water sweet. With plenty of light you will see, to quote Mr. Gosse once more, [33] "thousands of tiny globules forming on every plant, and even all over the stones, where the infant vegetation is beginning to grow; and these globules presently rise in rapid succession to the surface all over the vessel, and this process goes on uninterruptedly as long as the rays of the sun are uninterrupted.

"Now these globules consist of PURE OXYGEN, given out by the plants under the stimulus of light; and to this oxygen the animals in the tank owe their life. The difference between the profusion of oxygen-bubbles produced on a sunny day, and the paucity of those seen on a dark cloudy day, or in a northern aspect, is very marked." Choose, therefore, a south or east window, but draw down the blind, or throw a handkerchief over all if the heat become fierce. The water should always feel cold to your hand, let the temperature outside be what it may.

Next, you must make up for evaporation by FRESH water (a very little will suffice), as often as in summer you find the water in your vase sink below its original level, and prevent the water from getting too salt. For the salts, remember, do not evaporate with the water; and if you left the vase in the sun for a few weeks, it would become a mere brine-pan.

But how will you move your treasures up to town?

The simplest plan which I have found successful is an earthen jar. You may buy them with a cover which screws on with two iron clasps. If you do not find such, a piece of oilskin tied over the mouth is enough. But do not fill the jar full of water; leave about a quarter of the contents in empty air, which the water may absorb, and so keep itself fresh. And any pieces of stone, or oysters, which you send up, hang by a string from the mouth, that they may not hurt tender animals by rolling about the bottom. With these simple precautions, anything which you are likely to find will well endure forty-eight hours of travel.

What if the water fails, after all?

Then Mr. Gosse's artificial sea-water will form a perfect substitute. You may buy the requisite salts (for there are more salts than "salt" in sea-water) from any chemist to whom Mr. Gosse has entrusted his discovery, and, according to his directions, make sea-water for yourself

One more hint before we part. If, after all, you are not going down to the sea-side this year, and have no opportunities of testing "the wonders of the shore," you may still study Natural History in your own drawing-room, by looking a little into "the wonders of the pond."

I am not jesting; a fresh-water aquarium, though by no means as beautiful as a salt-water one, is even more easily established. A glass jar, floored with two or three inches of pond-mud (which should be covered with fine gravel to prevent the mud washing up); a specimen of each of two water-plants which you may buy now at any good shop in Covent Garden, Vallisneria spiralis (which is said to give to the Canvas-backed duck of America its peculiar richness of flavour), and Anacharis alsinastrum, that magical weed which, lately introduced from Canada among timber, has multiplied, self-sown, to so prodigious an extent, that it bid fair, a few years since, to choke the navigation not only of our canals and fen- rivers, but of the Thames itself:[34] or, in default of these, some of the more delicate pond-weeds; such as Callitriche, Potamogeton pusillum, and, best of all, perhaps, the beautiful Water-Milfoil (Myriophyllium), whose comb-like leaves are the haunts of numberless rare and curious animalcules:- these (in themselves, from the transparency of their circulation, interesting microscopic objects) for oxygen-breeding vegetables; and for animals, the pickings of any pond; a minnow or two, an eft; a few of the delicate pond-snails (unless they devour your plants too rapidly): water-beetles, of activity inconceivable, and that wondrous bug the Notonecta, who lies on his back all day, rowing about his boat-shaped body, with one long pair of oars, in search of animalcules, and the moment the lights are out, turns head over heels, rights himself, and opening a pair of handsome wings, starts to fly about the dark room in company with his friend the water- beetle, and (I suspect) catch flies; and then slips back demurely into the water with the first streak of dawn. But perhaps the most interesting of all the tribes of the Naiads, - (in default, of course, of those semi-human nymphs with which our Teutonic forefathers, like the Greeks, peopled each "sacred fountain,") - are the little "water-crickets," which may be found running under the pebbles, or burrowing in little galleries in the banks: and those "caddises," which crawl on the bottom in the stiller waters, enclosed, all save the head and legs, in a tube of sand or pebbles, shells or sticks, green or dead weeds, often arranged with quaint symmetry, or of very graceful shape. Their aspect in this state may be somewhat uninviting, but they compensate for their youthful ugliness by the strangeness of their transformations, and often by the delicate beauty of the perfect insects, as the "caddises," rising to the surface, become flying Phryganeae (caperers and sand- flies), generally of various shades of fawn-colour; and the water- crickets (though an unscientific eye may be able to discern but little difference in them in the "larva," or imperfect state) change into flies of the most various shapes; - one, perhaps, into the great sluggish olive "Stone-fly" (Perla bicaudata); another into the delicate lemon-coloured "Yellow Sally" (Chrysoperla viridis); another into the dark chocolate "Alder" (Sialis lutaria): and the majority into duns and drakes (Ephemerae); whose grace of form, and delicacy of colour, give them a right to rank among the most exquisite of God's creations, from the tiny "Spinners" (Batis or

Chloron) of incandescent glass, with gorgeous rainbow-coloured eyes, to the great Green Drake (Ephemera vulgata), known to all fishermen as the prince of trout-flies. These animals, their habits, their miraculous transformations, might give many an hour's quiet amusement to an invalid, laid on a sofa, or imprisoned in a sick-room, and debarred from reading, unless by some such means, any page of that great green book outside, whose pen is the finger of God, whose covers are the fire kingdoms and the star kingdoms, and its leaves the heather-bells, and the polypes of the sea, and the gnats above the summer stream.

I said just now, that happy was the sportsman who was also a naturalist. And, having once mentioned these curious water-flies, I cannot help going a little farther, and saying, that lucky is the fisherman who is also a naturalist. A fair scientific knowledge of the flies which he imitates, and of their habits, would often ensure him sport, while other men are going home with empty creels. One would have fancied this a self-evident fact; yet I have never found any sound knowledge of the natural water-flies which haunt a given stream, except among cunning old fishermen of the lower class, who get their living by the gentle art, and bring to indoors baskets of trout killed on flies, which look as if they had been tied with a pair of tongs, so rough and ungainly are they; but which, nevertheless, kill, simply because they are (in COLOUR, which is all that fish really care for) exact likenesses of some obscure local species, which happen to be on the water at the time. Among gentlemen-fishermen, on the other hand, so deep is the ignorance of the natural fly, that I have known good sportsmen still under the delusion that the great green May-fly comes out of a caddis-bait; the gentlemen having never seen, much less fished with, that most deadly bait the "Water-cricket," or free creeping larva of the May-fly, which may be found in May under the river-banks. The consequence of this ignorance is that they depend for good patterns of flies on mere chance and experiment; and that the shop patterns, originally excellent, deteriorate continually, till little or no likeness to their living prototype remains, being tied by town girls, who have no more understanding of what the feathers and mohair in their hands represent than they have of what the National Debt represents. Hence follows many a failure at the stream-side; because the "Caperer," or "Dun," or "Yellow Sally," which is produced from the fly-book, though, possibly, like the brood which came out three years since on some stream a hundred miles away, is quite unlike the brood which is out to-day on one's own river. For not only do most of these flies vary in colour in different soils and climates, but many of them change their hue during life; the Ephemerae, especially, have a habit of throwing off the whole of their skins (even, marvellously enough, to the skin of the eyes and wings, and the delicate "whisks" at their tail), and appearing in an utterly new garb after ten minutes' rest, to the discomfiture of the astonished angler.

The natural history of these flies, I understand from Mr. Stainton (one of our most distinguished entomologists), has not yet been worked out, at least for England. The only attempt, I believe, in that direction is one made by a charming book, "The Fly-fisher's Entomology," which should be in every good angler's library; but why should not a few fishermen combine to work out the subject for themselves, and study for the interests both of science and their own sport, "The Wonders of the Bank?" The work, petty as it may seem, is much too great for one man, so prodigal is Nature of her forms, in the stream as in the ocean; but what if a correspondence were opened between a few fishermen - of whom one should live, say, by the Hampshire or Berkshire chalk streams; another on the slates and granites of Devon; another on the limestones of Yorkshire or Derbyshire; another among the yet earlier slates of Snowdonia, or some mountain part of Wales; and more than one among the hills of the Border and the lakes of the Highlands? Each would find (I suspect), on comparing his insects with those of the others, that he was exploring a little peculiar world of his own, and that with the exception of a certain number of typical forms, the flies of his county were unknown a hundred miles away, or, at least, appeared there under great differences of size and colour; and each, if he would take the trouble to collect the caddises and water-crickets, and breed them into the perfect fly in an aquarium, would see marvels in their transformations, their instincts, their anatomy, quite as great (though not, perhaps, as showy and startling) as I have been trying to point out on the sea-shore. Moreover, each and every one of the party, I will warrant, will find his fellow-correspondents (perhaps previously unknown to him) men worth knowing; not, it may be, of the meditative and half- saintly type of dear old Izaak Walton (who, after all, was no fly- fisher, but a sedentary "popjoy" guilty of float and worm), but rather, like his fly-fishing disciple Cotton, good fellows and men of the world, and, perhaps, something better over and above.

The suggestion has been made. Will it ever be taken up, and a "Naiad Club" formed, for the combination of sport and science?

And, now, how can this desultory little treatise end more usefully than in recommending a few books on Natural History, fit for the use of young people; and fit to serve as introductions to such deeper and larger works as Yarrell's "Birds and Fishes," Bell's "Quadrupeds" and "Crustacea," Forbes and Hanley's "Mollusca," Owen's "Fossil Mammals and Birds," and a host of other admirable works? Not that this list will contain all the best; but simply the best of which the writer knows; let, therefore, none feel aggrieved, if, as it may chance, opening these pages, they find their books omitted.

First and foremost, certainly, come Mr. Gosse's books. There is a playful and genial spirit in them, a brilliant power of word-painting combined with deep and earnest religious feeling, which makes them as morally valuable as they are intellectually interesting. Since White's "History of Selborne," few or no writers on Natural History, save Mr. Gosse, Mr. G. H. Lewes, and poor Mr. E. Forbes, have had the power of bringing out the human side of science, and giving to seemingly dry disquisitions and animals of the lowest type, by little touches of pathos and humour, that living and personal interest, to bestow which is generally the special function of the poet: not that Waterton and Jesse are not excellent in this respect, and authors who should be in every boy's library: but they are rather anecdotists than systematic or scientific inquirers; while Mr. Gosse, in his "Naturalist on the Shores of Devon," his "Tour in Jamaica," his "Tenby," and his "Canadian Naturalist," has done for those three places what White did for Selborne, with all the improved appliances of a science which has widened and deepened tenfold since White's time. Mr. Gosse's "Manual of the Marine Zoology of the British Isles" is, for classification, by far the completest handbook extant. He has contrived in it to compress more sound knowledge of vast classes of the animal kingdom than I ever saw before in so small a space. [35]

Miss Anne Pratt's "Things of the Sea-coast" is excellent; and still better is Professor Harvey's "Sea-side Book," of which it is impossible to speak too highly; and most pleasant it is to see a man of genius and learning thus gathering the bloom of his varied knowledge, to put it into a form equally suited to a child and a SAVANT. Seldom, perhaps, has there been a little book in which so vast a quantity of facts have been told so gracefully, simply, without a taint of pedantry or cumbrousness - an excellence which is the sure and only mark of a perfect mastery of the subject. Mr. G. H. Lewes's "Sea-shore Studies" are also very valuable; hardly perhaps a book for beginners, but from his admirable power of description, whether of animals or of scenes, is interesting for all classes of readers.

Two little "Popular" Histories - one of British Zoophytes, the other of British Sea-weeds, by Dr. Landsborough (since dead of cholera, at Saltcoats, the scene of his energetic and pious ministry) - are very excellent; and are furnished, too, with well-drawn and coloured plates, for the comfort of those to whom a scientific nomenclature (as liable as any other human thing to be faulty and obscure) conveys but a vague conception of the objects. These may serve well for the beginner, as introductions to Professor Harvey's large work on British Algae, and to the new edition of Professor Johnston's invaluable "British Zoophytes," Miss Gifford's "Marine Botanist," third edition, and Dr. Cocks's "Sea-weed Collector's Guide," have also been recommended by a high authority.

For general Zoology the best books for beginners are, perhaps, as a general introduction, the Rev. J. A. L. Wood's "Popular Zoology," full of excellent plates; and for systematic Zoology, Mr. Gosse's four little books, on Mammals, Birds, Reptiles, and Fishes, published with many plates, by the Christian Knowledge Society, at a marvellously cheap rate. For miscroscopic animalcules, Miss Agnes Catlow's "Drops of Water" will teach the young more than they will ever remember, and serve as a good introduction to those teeming abysses of the unseen world, which must be afterwards traversed under the guidance of Hassall and Ehrenberg.

For Ornithology, there is no book, after all, like dear old Bewick, PASSE though he may be in a scientific point of view. There is a good little British ornithology, too, published in Sir W. Jardine's "Naturalist's Library," and another by Mr. Gosse. And Mr. Knox's "Ornithological Rambles in Sussex," with Mr. St. John's "Highland

Sports," and "Tour in Sutherlandshire," are the monographs of naturalists, gentlemen, and sportsmen, which remind one at every page (and what higher praise can one give?) of White's "History of Selborne." These last, with Mr. Gosse's "Canadian Naturalist," and his little book "The Ocean," not forgetting Darwin's delightful "Voyage of the Beagle and Adventure," ought to be in the hands of every lad who is likely to travel to our colonies.

For general Geology, Professor Ansted's Introduction is excellent; while, as a specimen of the way in which a single district may be thoroughly worked out, and the universal method of induction learnt from a narrow field of objects, what book can, or perhaps ever will, compare with Mr. Hugh Miller's "Old Red Sandstone"?

For this last reason, I especially recommend to the young the Rev. C. A. Johns's "Week at the Lizard," as teaching a young person how much there is to be seen and known within a few square miles of these British Isles. But, indeed, all Mr. Johns's books are good (as they are bound to be, considering his most accurate and varied knowledge), especially his "Flowers of the Field," the best cheap introduction to systematic botany which has yet appeared. Trained, and all but self-trained, like Mr. Hugh Miller, in a remote and narrow field of observation, Mr. Johns has developed himself into one of our most acute and persevering botanists, and has added many a new treasure to the Flora of these isles; and one person, at least, owes him a deep debt of gratitude for first lessons in scientific accuracy and patience, - lessons taught, not dully and dryly at the book and desk, but livingly and genially, in adventurous rambles over the bleak cliffs and ferny woods of the wild Atlantic shore, -

> "Where the old fable of the guarded mount
> Looks toward Namancos and Bayona's hold."

Mr. Henfrey's "Rudiments of Botany" might accompany Mr. Johns's books. Mr. Babington's "Manual of British Botany" is also most compact and highly finished, and seems the best work which I know of from which a student somewhat advanced in English botany can verify species; while for ferns, Moore's "Handbook" is probably the best for beginners.

For Entomology, which, after all, is the study most fit for boys (as Botany is for girls) who have no opportunity for visiting the sea-shore, Catlow's "Popular British Entomology," having coloured plates (a delight to young people), and saying something of all the orders, is, probably, still a good work for beginners.

Mr. Stainton's "Entomologist's Annual for 1855" contains valuable hints of that gentleman's on taking and arranging moths and butterflies; as well as of Mr. Wollaston's on performing the same kind office for that far more numerous, and not less beautiful class, the beetles. There is also an admirable "Manual of British Butterflies and Moths," by Mr. Stainton, in course of publication; but, perhaps, the most interesting of all entomological books which I have seen (and for introducing me to which I must express my hearty thanks to Mr. Stainton), is "Practical Hints respecting Moths and Butterflies, forming a Calendar of Entomological Operations," [36] by Richard Shield, a simple London working-man.

I would gladly devote more space than I can here spare to a review of this little book, so perfectly does it corroborate every word which I have said already as to the moral and intellectual value of such studies. Richard Shield, making himself a first-rate "lepidopterist," while working with his hands for a pound a week, is the antitype of Mr. Peach, the coast-guardsman, among his Cornish tide-rocks. But more than this, there is about Shield's book a tone as of Izaak Walton himself, which is very delightful; tender, poetical, and religious, yet full of quiet quaintness and humour; showing in every page how the love for Natural History is in him only one expression of a love for all things beautiful, and pure, and right. If any readers of these pages fancy that I over- praise the book, let them buy it, and judge for themselves. They will thus help the good man toward pursuing his studies with larger and better appliances, and will be (as I expect) surprised to find how much there is to be seen and done, even by a working-man, within a day's walk of smoky Babylon itself; and how easily a man might, if he would, wash his soul clean for a while from all the turmoil and intrigue, the vanity and vexation of spirit of that "too-populous wilderness," by going out to be alone a while with God in heaven, and with that earth which He has given to the children of men, not merely for the material wants of their bodies, but as a witness and a sacrament that in Him they live and move, and have their being, "not by bread alone, but by EVERY word that proceedeth out of the mouth of God."

Thus I wrote some twenty years ago, when the study of Natural History was confined mainly to several scientific men, or mere collectors of shells, insects, and dried plants.

Since then, I am glad to say, it has become a popular and common pursuit, owing, I doubt not, to the impulse given to it by the many authors whose works I then recommended. I recommend them still; though a swarm of other manuals and popular works have appeared since, excellent in their way, and almost beyond counting. But all honour to those, and above all to Mr. Gosse and Mr. Johns, who first opened people's eyes to the wonders around them all day long. Now, we have, in addition to amusing books on special subjects, serials on Natural History more or less profound, and suited to every kind of student and every grade of knowledge. I mention the names of none. For first, they happily need no advertisement from me; and next, I fear to be unjust to any one of them by inadvertently omitting its name. Let me add, that in the advertising columns of those serials, will be found notices of all the new manuals, and of all apparatus, and other matters, needed by amateur naturalists, and of many who are more than amateurs. Microscopy, meanwhile, and the whole study of "The Wonders of the Little," have made vast strides in the last twenty years; and I was equally surprised and pleased, to find, three years ago, in each of two towns of a few thousand inhabitants, perhaps a dozen good microscopes, all but hidden away from the public, worked by men who knew how to handle them, and who knew what they were looking at; but who modestly refrained from telling anybody what they were doing so well. And it was this very discovery of unsuspected microscopists which made me more desirous than ever to see - as I see now in many places - scientific societies, by means of which the few, who otherwise would work apart, may communicate their knowledge to each other, and to the many. These "Microscopic," "Naturalist," "Geological," or other societies, and the "Field Clubs" for excursions into the country, which are usually connected with them, form a most pleasant and hopeful new feature in English Society; bringing together, as they do, almost all ranks, all shades of opinion; and it has given me deep pleasure to see, in the case at least of the Country Clubs with which I am acquainted, the clergy of the Church of England taking an active, and often a leading, interest in their practical work. The town clergy are, for the most part, too utterly overworked to follow the example of their country brethren. But I have reason to know that they regard such societies, and Natural History in general, with no unfriendly eyes; and that there is less fear than ever that the clergy of the Church of England should have to relinquish their ancient boast - that since the formation of the Royal Society in the seventeenth century, they have done more for sound physical science than any other priesthood or ministry in the world. Let me advise anyone who may do me the honour of reading these pages, to discover whether such a Club or Society exists in his neighbourhood, and to join it forthwith, certain that - if his experience be at all like mine - he will gain most pleasant information and most pleasant acquaintances, and pass most pleasant days and evenings, among people whom he will be glad to know, and whom he never would have known save for the new - and now, I hope, rapidly spreading - freemasonry of Natural History.

Meanwhile, I hope - though I dare not say I trust - to see the day when the boys of each of our large schools shall join - like those of Marlborough and Clifton - the same freemasonry; and have their own Naturalists' Clubs; nay more; when our public schools and universities shall awake to the real needs of the age, and - even to the curtailing of the time usually spent in not learning Latin and Greek - teach boys the rudiments at least of botany, zoology, geology, and so forth; and when the public opinion, at least of the refined and educated, shall consider it as ludicrous - to use no stronger word - to be ignorant of the commonest facts and laws of this living planet, as to be ignorant of the rudiments of two dead languages. All honour to the said two languages. Ignorance of them is a serious weakness; for it implies ignorance of many things else; and indeed, without some knowledge of them, the nomenclature of the physical sciences cannot be mastered. But I have got to discover that a boy's time is more usefully spent, and his intellect more methodically trained, by getting up Ovid's Fasti with an ulterior hope of being able to write a few Latin verses, than in getting up Professor Rolleston's "Forms of Animal Life," or any other of the excellent Scientific Manuals for beginners, which are now, as I said, happily so numerous.

May that day soon come; and an old dream of mine, and of my scientific friends, be fulfilled at last.

And so I end this little book, hoping, even praying, that it may encourage a few more labourers to go forth into a vineyard, which those who have toiled in it know to be full of ever-fresh health, and wonder and simple joy, and the presence and the glory of Him whose name is LOVE.

APPENDIX.

PLATE I.

ZOOPHYTA. POLYZOA.

THE forms of animal life which are now united in an independent class, under the name Polyzoa, so nearly resemble the Hydroid Zoophytes in general form and appearance that a casual observer may suppose them to be nearly identical. In all but the more recent works, they are treated as distinct indeed, but still included under the general term "ZOOPHYTES." The animals of both groups are minute, polypiform creatures, mostly living in transparent cells, springing from the sides of a stem which unites a number of individuals in one common life, and grows in a shrub-like form upon any submarine body, such as a shell, a rock, a weed, or even another polypidom to which it is parasitically attached. Each polype, in both classes, protrudes from and retreats within its cell by an independent action, and when protruded puts forth a circle of tentacles whose motion round the mouth is the means of securing nourishment. There are, however, peculiarities in the structure of the Polyzoa which seem to remove them from Zoophytology to a place in the system of nature more nearly connected with Molluscan types. Some of them come so near to the compound ascidians that they have been termed, as an order, "Zoophyta ascidioida."

The simplest form of polype is that of a fleshy bag open at one end, surmounted by a circle of contractile threads or fingers called tentacles. The plate shows, on a very minute scale, at figs. 1, 3, and 6, several of these little polypiform bodies protruding from their cells. But the Hydra or Fresh-water Polype has no cell, and is quite unconnected with any root thread, or with other individuals of the same species. It is perfectly free, and so simple in its structure, that when the sac which forms its body is turned inside out it will continue to perform the functions of life as before. The greater part, however, of these Hydraform Polypes, although equally simple as individuals, are connected in a compound life by means of their variously formed POLYPIDOM, as the branched system of cells is termed. The Hydroid Zoophytes are represented in the first plate by the following examples.

HYDROIDA.

SERTULARIA ROSEA. PL. I. FIG. 6.

A species which has the cells in pairs on opposite sides of the central tube, with the openings turned outwards. In the more enlarged figure is seen a septum across the inner part of each cell which forms the base upon which the polype rests. Fig. 6 B indicates the natural size of the piece of branch represented; but it must be remembered that this is only a small portion of the bushy shrub.

CAMPANULARIA SYRINGA. PL. I. FIG. 8.

This Zoophyte twines itself parasitically upon a species of Sertularia. The cells in this species are thrown out at irregular intervals upon flexible stems which are wrinkled in rings. They consist of lengthened, cylindrical, transparent vases.

CAMPANULARIA VOLUBILIS. PL. I. FIG. 9.

A still more beautiful species, with lengthened foot-stalks ringed at each end. The polype is remarkable for the protrusion and contractile power of its lips. It has about twenty knobbed tentacula.

POLYZOA.

Among Polyzoa the animal's body is coated with a membraneous covering, like that of the Tunicated Mollusca, but which is a continuation of the edge of the cell, which doubles back upon the body in such a manner that when the animal protrudes from its cell it pushes out the flexible membrane just as one would turn inside out the finger of a glove. This oneness of cell and polype is a distinctive character of the group. Another is the higher organization of the internal parts. The mouth, surrounded by tentacles, leads by gullet and gizzard through a channel into a digesting stomach, from which the rejectable matter passes upwards through an intestinal canal till it is discharged near the mouth.

The tentacles also differ much from those of true Polypes. Instead of being fleshy and contractile, they are rather stiff, resembling spun glass, set on the sides with vibrating cilia, which by their motion up one side and down the other of each tentacle, produce a current which impels their living food into the mouth. When these tentacles are withdrawn, they are gathered up in a bundle, like the stays of an umbrella. Our Plate I. contains the following examples of Polyzoa.

VALKERIA CUSCUTA. PL. I. FIG. 3.

From a group in one of Mr. Lloyd's vases. Fig. 3 A is the natural size of the central group of cells, in a specimen coiled round a thread-like weed. Underneath this is the same portion enlarged.

When magnified to this apparent size, the cells could be seen in different states, some closed, and others with their bodies protruded. When magnified to 3 D, we could pleasantly watch the gradual eversion of the membrane, then the points of the tentacles slowly appearing, and then, when fully protruded, suddenly expanding into a bell-shaped circle. This was their usual appearance, but sometimes they could be noticed bending inwards, as in fig. 3 C, as if to imprison some living atom of importance.

Fig. B represents two tentacles, showing the direction in which the cilia vibrate.

CRISIA DENTICULATA. PL. I. FIG. 4.

I have only drawn the cells from a prepared specimen. The polypes are like those described above.

GEMELLARIA LORICATA. PL. I. FIG. 5.

Here the cells are placed in pairs, back to back. 5 A is a very small portion on the natural scale.

CELLULARIA CILIATA. Pl. I. FIG. 7

The cells are alternate on the stem, and are curiously armed with long whip-like cilia or spines. On the back of some of the cells is a very strange appendage, the use of which is not with certainty ascertained. It is a minute body, slightly resembling a vulture's head, with a movable lower beak. The whole head keeps up a nodding motion, and the movable beak occasionally opens widely, and then suddenly snaps to with a jerk. It has been seen to hold an animalcule between its jaws till the latter has died, but it has no power to communicate the prey to the polype in its cell or to swallow and digest it on its own account. It is certainly not an independent parasite, as has been supposed, and yet its purpose in the animal economy is a mystery. Mr. Gosse conjectures that its use may be, by holding animalcules till they die and decay, to attract by their putrescence crowds of other animalcules, which may thus be drawn within the influence of the polype's ciliated tentacles. Fig. 7 B shows the form of one of these "birds' heads," and fig. 7 C, its position on the cell.

FLUSTRA LINEATA. PL. I. FIG. 1.

In Flustrae, the cells are placed side by side on an expanded membrane. Fig. 1 represents the general appearance of a species which at least resembles F. lineata as figured in Johnston's work. It is spread upon a Fucus. Fig. A is an enlarged view of the cells.

FLUSTRA FOLIACEA. PL. I. FIG. 2.

We figure a frond or two of the common species, which has cells on both sides. It is rarely that the polypes can be seen in a state of expansion.

SERIALARIA LENDIGERA. PL. I. fig. 10.
NOTAMIA BURSARIA. PL. I. fig. 11.

The "tobacco-pipe"" appendages, fig. 11 B, are of unknown use: they are probably analogous to the birds' heads in the Cellularae.

PLATE V.

CORALS AND SEA ANEMONES.

CARYOPHYLLAEA SMITHII. PL. V. FIG. 2. PL. VI. FIG. 3.

THE connection between Brainstones, Mushroom Corals, and other Madrepores abounding on Polynesian reefs, and the "Sea Anemones," which have lately become so familiar to us all, can be seen by comparing our comparatively insignificant C. Smithii with our commonest species of Actinia and Sagartia. The former is a beautiful object when the fleshy part and tentacles are wholly or partially expanded. Like Actinia, it has a membranous covering, a simple sac-like stomach, a central mouth, a disk surrounded by contractile and adhesive tentacles. Unlike Actinia, it is fixed to submarine bodies, to which it is glued in very early life, and cannot change its place. Unlike Actinia, its body is supported by a stony skeleton of calcareous plates arranged edgewise so as to radiate from the centre. But as we find some Molluscs furnished with a shell, and others even of the same character and habits without one, so we find that in spite of this seemingly important difference, the animals are very similar in their nature. Since the introduction of glass tanks we have opportunities of seeing anemones crawling up the sides, so as to exhibit their entire basal disk, and then we may observe lightly coloured lines of a less transparent substance than the interstices, radiating from the margin to the centre, some short, others reaching the entire distance, and arranged in exactly the same manner as the plates of Caryophyllaea. These are doubtless flexible walls of compartments dividing the fleshy parts of the softer animals, and corresponding with the septa of the coral. Fig.

2 A represents a section of the latter, to be compared with the basal disk of Sagartia.

SAGARTIA ANGUICOMA. PL. V. FIG. 3, A, B.

This genus has been separated from Actinia on account of its habit of throwing out threads when irritated. Although my specimens often assumed the form represented in fig. 3, Mr. Lloyd informs me that it must have arisen from unhealthiness of condition, its usual habit being to contract into a more flattened form. When fully expanded, its transparent and lengthened tentacles present a beautiful appearance. Fig. 3 A, showing a basal disk, is given for the purpose already described.

BALANOPHYLLAEA REGIA. PL. V. FIG. 1.

Another species of British madrepore, found by Mr. Gosse at Ilfracombe, and by Mr. Kingsley at Lundy Island. It is smaller than O. Smithii, of a very bright colour, and always covers the upper part of its bony skeleton, in which the plates are differently arranged from those of the smaller species. Fig. 1 shows the tentacles expanded in an unusual degree; 1 A, animal contracted; 1 B, the coral; 1 C, a tentacle enlarged.

PLATE VI.

CORALS AND SEA ANEMONES.

ACTINIA MESEMBRYANTHEMUM. PL. VI. FIG. 1 A.

This common species is more frequently met with than many others, because it prefers shallow water, and often lives high up among rocks which are only covered by the sea at very high tide; so that the creature can, if it will, spend but a short portion of its time immersed. When uncovered by the tide, it gathers up its leathery tunic, and presents the appearance of fig. 1 A. When under water it may often be seen expanding its flower-like disk and moving its feelers in search of food. These feelers have a certain power of adhesion, and any not too vigorous animals which they touch are easily drawn towards the centre and swallowed. Around the margin of the tunic are seen peeping out between the tentacles certain bright blue globules looking very like eyes, but whose purpose is not exactly ascertained. Fig. 1 represents the disk only partially expanded.

BUNODES CRASSICORNIS. PL. VI. FIG. 2.

This genus of Actinioid zoophytes is distinguished from Actinia proper by the tubercles or warts which stud the outer covering of the animal. In B. gemmacea these warts are arranged symmetrically, so as to give a peculiarly jewelled appearance to the body. Being of a large size, the tentacles of B. crassicornis exhibit in great perfection the adhesive powers produced by the nettling threads which proceed from them.

CARYOPHYLLAEA SMITHII. PL. VI. FIG. 3.

This figure is to show a whiter variety, with the flesh and tentacles fully expanded

PLATE VIII.

MOLLUSCA.

NASSA RETICULATA. PL. VIII. fig. 2, A, B, C, D, E, F

A VERY active Mollusc, given here chiefly on account of the opportunity afforded by the birth of young fry in Mr. Lloyd's tanks. The NASSA feeds on small animalcules, for which, in aquaria, it may be seen routing among the sand and stones, sometimes burying itself among them so as only to show its caudal tube moving along between them. A pair of Nassae in Mr. Lloyd's collection, deposited, on the 5th of April, about fifty capsules or bags of eggs upon the stems of weeds (fig. 2 B); each capsule contained about a hundred eggs. The capsules opened on the 16th of May, permitting the escape of rotiferous fry (fig. 2, C, D, E), not in the slightest degree resembling the parent, but presenting minute nautilus-shaped transparent shells. These shells rather hang on than cover the bodies, which have a pair of lobes, around which vibrate minute cilia in such a manner as to give them an appearance of rotatory motion. Under a lens they may be seen moving about very actively in various positions, but always with the look of being moved by rapidly turning wheels. We should have been glad to witness the next step towards assuming their ultimate form, but were disappointed, as the embryos died. Fig. 2 F is the tongue of a Nassa, from a photograph by Dr. Kingsley.

FOOTNOTES:

1. SERTULARIA OPERCULATA and GEMELLARIA LOCICULATA; or any of the small SERTULARIAE, compared with CRISIAE and CELLULARIAE, are very good examples. For a fuller description of these, see Appendix explaining Plate I.
2. If any inland reader wishes to see the action of this foot, in the bivalve Molluscs, let him look at the Common Pond-Mussel (Anodon Cygneus), which he will find in most stagnant waters, and see how he burrows with it in the mud, and how, when the water is drawn off, he walks solemnly into deeper water, leaving a furrow behind him.
3. These shells are so common that I have not cared to figure them.
4. Plate IX. Fig. 3, represents both parasites on the dead Turritella.
5. A few words on him, and on sea-anemones in general, may be found in Appendix II. But full details, accompanied with beautiful plates, may be found in Mr. Gosse's work on British sea-anemones and madrepores, which ought to be in every seaside library.
6. Handbook to the Marine Aquarium of the Crystal Palace.
7. An admirable paper on this extraordinary family may be found in the Zoological Society's Proceedings for July 1858, by Messrs. S. P. Woodward and the late lamented Lucas Barrett. See also Quatrefages, I. 82, or Synapta Duvernaei.
8. Thalassema Neptuni (Forbes' British Star-Fishes, p. 259),
9. The Londoner may see specimens of them at the Zoological Gardens and at the Crystal Palace; as also of the rare and beautiful Sabella, figured in the same plate; and of the Balanophyllia, or a closely-allied species, from the Mediterranean, mentioned in p. 109.
10. A Naturalist's Rambles on the Devonshire Coast, p. 110.
11. Balanophyllia regia, Plate V. fig. 1.
12. Amphidotus cordatus.
13. Echinus miliaris, Plate VII.

14　See Professor Sedgwick's last edition of the "Discourses on the Studies of Cambridge."
15　Fissurella graeca, Plate X. fig. 5.
16　Doris tuberculata and bilineata.
17　Eolis papi losa. A Doris and an Eolis, though not of these species, are figured in Plate X.
18　Plate III.
19　Certain Parisian zoologists have done me the honour to hint that this description was a play of fancy. I can only answer, that I saw it with my own eyes in my own aquarium. I am not, I hope, in the habit of drawing on my fancy in the presence of infinitely more marvellous Nature. Truth is quite strange enough to be interesting without lies.
20　Saxicava rugosa, Plate XI. fig. 2.
21　Plate VIII. represents the common Nassa, with the still more common Littorina littorea, their teeth-studded palates, and the free swimming young of the Nassa. (VIDE Appendix.)
22　Cyproea Europoea.
23　Botrylli.
24　Molluscs.
Doris tuberculata.
- bilineata.
Eolis papillosa.
Pleurobranchus plumila.
Neritina.
Cypraea.
Trochus, - 2 species.
Mangelia.
Triton.
Trophon.
Nassa, - 2 species.
Cerithium.
Sigaretus.
Fissurella.
Arca lactea.
Pecten pusio.
Tapes pullastra.
Kellia suborbicularis.
Shaenia Binghami.
Saxicava rugosa.

Gastrochoena pholadia.
Pholas parva.
Anomiae, -2 or 3 species
Cynthia,-2 species.
Botryllus, do.
ANNELIDS.
Phyllodoce, and other Nereid worms.
Polynoe squamata.
CRUSTACEA.
4 or 5 species.
ECHINODERMS.
Echinus miliaris.
Asterias gibbosa.
Ophiocoma neglecla.
Cucumaria Hyndmanni.
- communis.
POLYPES.
Sertularia pumila.
- rugosa.
- fallax.
- filicula.
Plumularia falcata.
- setacea.
Laomedea geniculata.
Campanularia volubilis.
Actinia mesembryanthemum.
Actinia clavata.
- anguicoma.
- crassicornis.
Tubulipora patina.
- hispida.
- serpens.
Crisia eburnea.
Cellepora pumicosa.
Lepraliae,- many species.
Membranipora pilosa.
Cellularia ciliata.
- scruposa.
- reptans.
Flustra membranacea, &c.

25 Plate XI. fig. 1.
26 Plate X. fig. 1.
27 There are very fine specimens in the Crystal Palace.
28 Coryne ramosa.
29 Campanularia integra.
30 Crisidia Eburnea.
31 Aquarium, p. 163.
32 P. 34. Figures of it are given in Plate VIII.
33 P. 259.
34 But if any young lady, her aquarium having failed, shall (as dozens do) cast out the same Anacharis into the nearest ditch, she shall be followed to her grave by the maledictions of all millers and trout-fishers. Seriously, this is a wanton act of injury to the neighbouring streams, which must be carefully guarded against.

    As well turn loose queen-wasps to build in your neighbour's banks.
35 Very highly also, in interest, ranks M. Quatrefages' "Rambles of a Naturalist" (about the Mediterranean and the French Coast), translated by M. Otte.
36 Van Voorst & Co. price 3s.

BIBLIOBAZAAR

The essential book market!

Did you know that you can get any of our titles in large print?

Did you know that we have an ever-growing collection of books in many languages?

Order online:
www.bibliobazaar.com

Find all of your favorite classic books!

Stay up to date with the latest government reports!

At BiblioBazaar, we aim to make knowledge more accessible by making thousands of titles available to you- *quickly and affordably*.

Contact us:
BiblioBazaar
PO Box 21206
Charleston, SC 29413